Land of the Arabs

M. Abdel-Kader Hatem

Supervisor General of the Specialised National
Councils of Egypt, and formerly Deputy
Prime Minister of Egypt

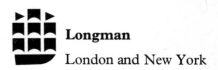

Longman
London and New York

0527724

LONGMAN GROUP LTD
London and New York

*Associated companies, branches and representatives
throughout the world.*

© Longman Group Ltd 1977

First published 1977

ISBN 0 582 78037 3

Library of Congress Cataloging in Publication Data

Hātim, Muhammad 'Abd al-Qādir.
  Land of the Arabs.

  Includes index.
  1. Arab countries.   I. Title.
DS36.7.H37   909'.09'74927   77-23164
ISBN 0-582-78037-3

Printed in Britain by W. S. Cowell Ltd, Ipswich

Dedication

This is a book which I have always wanted to write, for, as a professional information man, I have been astounded at the lack of knowledge in the Western world of the Arabs generally and of their land in particular.

To say that this is distressing to Arabs everywhere is putting it very mildly indeed, for the Arabs are a proud people: proud of their history, proud of their land, proud of their culture, and proud of their whole heritage.

This book is an attempt at making some of the basic physical facts about the Land of the Arabs known to the outside world.

It is to all my fellow Arabs who share with me the intense desire to tell the world about our land and our people that I dedicate this book.

Preface

The aim of this book is to give an account of the efforts of the Arabs past and present to develop their huge territory and to make the best use of the gifts that God has given them in overcoming the problems with which the nature of that land has always confronted them.

I hope that the book will be of interest and perhaps of value, not only to the Western reader who is already knowledgeable about the Arab World but to the vast majority of Europeans and Americans whose ignorance of the Arabs and Arab countries is profound. It is still perfectly possible for Westerners educated in some of the greatest universities of the world to emerge without the slightest idea of the Arabs' contribution to Western culture or indeed of the very nature of the Arab World; and if the intellectual élite find themselves so ignorant, through no fault of their own, it cannot be surprising if the average Westerner still thinks of Arabs in terms of rigid and inaccurate stereotypes, if his mental image of their countries consists of an endless panorama of sand, and if he feels in his heart of hearts that the Arabs are really a rather backward and curious lot.

Obviously, as its title makes clear, this book does not attempt to cover more than one aspect of the Arab World. But this is an area where Western ignorance is peculiarly profound and one of which Israel and Zionist propagandists have taken the fullest advantage; notably in planting in Western minds the totally false notion that before the creation of Israel in 1948 Palestine consisted of nothing but rock, sand and scrub, and that it is only the Israelis who have even attempted, let alone achieved, the task of 'making the desert flower'.

I have therefore tried to show, in varying detail, how false this picture is, not only in Palestine but throughout the Arab World. I am not an academic geographer, nor can I claim to be an irrigation engineer, a qualified agriculturist or a specialist in any of the manifold disciplines touched upon in this book. But it has been my privilege to have played a part in the recent development of my own country of Egypt and to have been in regular and frequent contact with leaders of other Arab nations, and thus to be fortuitously in a position to survey, however inadequately, the whole of that vast area known as the Arab World.

It will be obvious to the reader that considerable disparities exist in the book between the treatments, both in length and in depth, given to various Arab nations. These differences arise from variations in my own personal knowledge, from the degree of relevance of the subject-matter of the book to a particular Arab country,

7

from the antiquity or otherwise of irrigation and reclamation practices which vary enormously from one country to another, and from the availability of records and data, past and present. Problems of land-ownership differ greatly from one Arab State to another, so that in some chapters much space is devoted to the ownership question while in others the topic is ignored. My Arab readers will therefore, I hope, forgive me if the space devoted to their own country seems slight compared with that allotted to others. There are reasons, or a combination of reasons, for disparities in depth and extent of treatment, and there is of course no question of my making value-judgements about the comparative 'importance' of any of the Arab States.

For the benefit of Western readers, I must explain why no chapter is devoted to Israel. The territory of Israel is dealt with in the chapters on Palestine and Jordan, and if my account of that territory seems to be less detailed than the area warrants, I make no apology, because the Israelis and their Zionist supporters have made sure that their efforts in this direction have been more than adequately made known to the world at large, usually with extreme shrillness and often with a high degree of exaggeration and inaccuracy. The book goes into a full explanation of my view of this whole contentious topic in the historical excursus on pages 222 to 226.

It remains for me to acknowledge with the greatest pleasure the enormous help that I have received from individuals, organisations and printed sources in the compilation and publication of this book. A comprehensive list of these would be almost as long as the book itself, but I should like in particular to express my indebtedness to that invaluable annual reference work *The Middle East and North Africa* (Europa Publications, London) and to other works of reference including *The United Arab Emirates* by K. G. Fenelon (Longman, London). I wish particularly to acknowledge the assistance I have unstintingly received from my publishers, Longman Group Limited. Mr Roger Stacey and Mr Jack Adam have personally worked with me on the editing of the text; while Mrs Janet Brown, designer and art-editor, has also made an invaluable contribution. To all these and to their colleagues in the Longman Arab World Division, I express my most grateful thanks.

M. Abdel-Kader Hatem

8

Contents

Preface   7
Introduction   13

The Arab Republic of Egypt   21
The Democratic Republic of the Sudan   87
The Libyan Arab Republic   114
The Republic of Tunisia   134
The Democratic Republic of Algeria   149
The Kingdom of Morocco   166
The Islamic Republic of Mauritania   182
The Republic of Lebanon   187
The Syrian Arab Republic   201
Palestine   222
The Hashemite Kingdom of Jordan   230
The Republic of Iraq   248
The Arabian Peninsula   271
The Kingdom of Saudi Arabia   273

The Gulf States
    The State of Kuwait   286
    The State of Bahrain   290
    The State of Qatar   290
    The United Arab Emirates   294

The Sultanate of Oman   296
The Arab Republic of Yemen (Northern Yemen)   299
The People's Democratic Republic of Yemen (Southern Yemen)   303
The Somali Democratic Republic   307

Index   313

Acknowledgements

We are grateful to the following for permission to reproduce copyright material:

American Elsevier Publishing Company, Inc., for diagrams from *The Agricultural Potential of the Middle East* by Marion Clawson, Hans M. Lansberg and Lyle T. Alexander; Central Agency for Public Mobilisation and Statistics, Egypt, for data from *Annual Statistical Abstract 1952–1966*, Cairo, June 1967; Europa Publications Limited for data from *The Middle East and North Africa 1975–76* (22nd Edition); Professor W. B. Fisher for diagrams from *The Middle East* (6th Edition); Department of Statistics, Jordan, for data from *Report on Agricultural Census*, 1965, June 1967 and *Report on the Results of the Agricultural Sample Survey*, 1966; Data by Farah Hassan Adam from *Research Bulletin No. 5* (1966), University of Khartoum;

Whilst every effort has been made we have been unable to trace copyright holders of the following and any information which would enable us to do so would be appreciated.
Data from *The Unified Development of the Water Resources of the Jordan Valley Region*, Boston 1953; data by Harold Hurst, Robert Black and Yusuf Samaika, *Nile Basin*, Vol. X, Cairo (1965);

The publishers would like to thank the following for the kindness and help which they gave during research for this book:
Messrs Coode and Partners, Consulting Engineers, London; Sir M. MacDonald and Partners, Consulting Engineers, London and Cambridge (particularly for their provision of data related to Sudan and Jordan); The Royal Geographical Society, London; Hunting Surveys Ltd, Boreham Wood; the staff of the East Sussex Library.

For permission to reproduce the photographs on the pages mentioned below the publishers are indebted to the following:
Aerofilms Ltd 174;
Associate Press Ltd 27, 80 (bottom);
Barnaby's Picture Library 137 (bottom), 141, 164 (top), 278 (bottom);
Camera Press Ltd 184 (both);
Christobel Grau 233 (top right);
Embassy of The Arab Republic of Egypt 42;
Embassy of The United Arab Emirates 293 (whole page);
Esso Petroleum Company Ltd 17 (top), 117 (both), 118 (both), 119 (bottom), 120, 121, 122, 124, 125, 126 (both), 127;
Herbert Addison, O.B.E., M.Sc., C.Eng. 49;
Hunting Technical Services Ltd 61, 89, 96, 112, 113, 154, 156, 234, 250, 253 (left), 278 (top);
Lebanese Tourist Authority 193;

Iraq Petroleum Company Ltd 194 (both), 195 (bottom), 211, 212, 214, 249, 255 (both), 267, 268, 288;

Middle East Archive 36 (both), 58 (bottom), 69, 73 (both), 233 (bottom), 281, 297 (both), 300 (bottom);

J. Allen Cash Ltd 26;

David Jameson Associates Ltd for The Ministry of Information and National Guidance, Somalia 310 (both);

National Aeronautics and Space Administration 24;

Paul Popper Ltd 33 (both), 91, 140 (bottom), 167, 168, 170 (both), 171, 175, 176 (both), 179, 233 (top left), 253 (right);

Picture Point Ltd 14, 140 (both top), 161 (whole page), 164 (bottom), 172 (both), 173, 185, 216 (top), 262, 276 (both), 289 (both);

Qatar Petroleum Company Ltd 291 (both);

Radio Times Hulton Picture Library 35, 63 (bottom), 65, 66 (both), 137 (top), 139, 143, 144 (both), 150;

Rex Features 84, 85 (both);

Royal Geographical Society 28, 43 (bottom), 63 (top), 118 (top), 234, 266;

Robert Harding Associates 17 (bottom), 46, 48, 58 (top), 70 (bottom), 205, 216 (bottom), 300 (both top);

Sir M. MacDonald and Partners 100 (whole page), 102, 228 (middle and bottom);

United Press International (U.K.) Ltd 38.

The Author.

Introduction

The Land of the Arabs is a vast territory of 13 million square kilometres, of which 28 per cent lies in Asia and 72 per cent in Africa. It stretches from the Atlantic coast of Morocco in the far west to the Arabian Sea in the east, and from Syria and Iraq in the north to the Sudan in the south.

Despite its great size, the Land of the Arabs forms a single geographical entity, surrounded as it is by natural boundaries—the Atlantic to the west, the Mediterranean and the Taurus range to the north, the Zagros Mountains, the Arabian Gulf, and the Arabian Sea to the east, and the Sahara Desert to the south. About 85 per cent of this enormous area consists of desert. Most of it is made up of a fairly high plateau, the surface of which has been levelled by erosion; so that, although parts are covered with light shallow soil, much of it is utterly bare of anything save rock and sand.

The Land of the Arabs consists, in fact, of a vast tract of mountain and desert, broken intermittently by valleys and oases. It is against this generally harsh environment that the Arabs have had to battle throughout their long history. The struggle still goes on, with more and more land reclaimed from the hostile desert as the talents and energies of modern Arab engineers are harnessed to the growing resources of technology, so that their achievements have become increasingly notable and numerous in recent times.

★   ★   ★

It is generally accepted that, of the few areas on earth where agriculture had its origin and gave rise to civilisation, three were in Arab lands: in the Nile Valley, on the plain of the Tigris and the Euphrates, and by the Marib Dam in Yemen.

In Egypt, for example, every condition favourable to agricultural development was present. The Nile flood arrived during the summer months, so that after the flood-waters had retreated, seeds could be sown early in the winter season. They thus had time to grow to maturity before they could be scorched by the heat of the sun in the following summer. As time went on, more and more of the fertile land was employed for the cultivation of barley, wheat and flax. Ancient Egyptian civilisation reached its first peak in the era of the Pyramid-builders (2680–2180 BC), and perhaps the most significant aspect of that civilisation was its complete dependence on the sons of Egypt alone, and on their use of Egypt's own natural resources, for agriculture.

13

Egypt was one of the areas of the world where agriculture originated.

On the plains of the Tigris and the Euphrates, too, a civilisation similar to that of Egypt arose, which depended on the waters of these two great rivers and on the fertile soil along their banks. In Yemen, the Marib Dam, designed to store rain-water, was a stupendous engineering feat which served to irrigate the land and thus led to the foundation of an ancient civilisation that earned Yemen the name 'Arabia Felix'. After the bursting of the great dam, this civilisation soon withered away.

Archaeologists are, of course, constantly making fresh discoveries and speculations about the various civilisations that sprang up on Arab soil. All agree that every one of these early civilisations was based upon the intelligent development of agriculture. At the end of the eighteenth century, French scientists who accompanied Napoleon on his Egyptian campaign studied in detail the methods then in use by Egyptian farmers. They found that their methods of raising water, which had certainly been in use for several millennia, were more efficient than those of the contemporary European farmer.

They noted also the simplicity and the effectiveness of the implements used by the Egyptian farmers for cultivation. Yet later researches show, from the study of early Egyptian pictorial art, that implements of an identical pattern were in use during the Pharaonic dynasties, and these have since formed the basis of that modern agricultural machinery for which electricity and the internal combustion engine provide the power.

Moreover, the Egyptians were the first to systematise irrigation, using elaborate and complex feats of engineering. They built dams and embankments to control the waters of the Nile, and dug the canal now known as Bahr Yusef to bring fertility to an area of desert.

The greatest irrigation scheme of the ancient world, the Marib Dam, was constructed in the far south of the Arabian Peninsula. Elsewhere, throughout the vast expanse of the Land of the Arabs, agriculture has been carried on from very early times in countless scattered oases dependent entirely on water drawn from artesian wells.

★　　★　　★

The Holy Quran describes Mecca as situated in an infertile valley, to remind us that the message of Islam sprang from a rough, stony, barren spot. Yet despite the fact that so much of the Land of the Arabs is desert, the Arabs' genius for agriculture has enabled them to exploit every area of fertile land and to extend it into previously uncultivable areas.

Their struggle against the sands of the desert has always depended on the greatest ally of all: water. The adventure stories told by desert explorers usually reach a happy ending when the thirsty caravan at last reaches a well brimming with sweet water. During his trek across Arabia and the Sinai Desert in the First World War, T. E. Lawrence was kept going only by the thought of a glass of iced lemonade awaiting him in the cool hall of the Heliopolis Palace Hotel on the outskirts of Cairo.

So the common belief of Europeans and Americans that the Arab countries consist only of a series of oases scattered through endless deserts is not entirely without foundation, although this view has often led foreign scholars into error and exaggeration. Fundamentally, the story of the Arabs through the ages has indeed been one of a constant search for water in desert places, and their great cultures and civilisations have been built around the oases and rivers of their land.

The Holy Quran says that water is the source of all life. There is also a reference to the two kinds of water: the sweet and delectable, and the pungent and bitter; Egypt is described as 'a paradise with sweet-water rivers running beneath it'. Water, and the crops to which water brings strength and life, feature prominently in the incidents and parables of the Holy Books which hand down to later generations the messages that have descended from Heaven on to the Land of the Arabs. The tale of Joseph's sojourn in Egypt is a story of agricultural vicissitudes; and when Moses led the Jews out of Egypt in flight from the Pharaoh, he sat by a well and watered the sheep for the girls who had been driven away by the shepherds. When Jesus came to Egypt with his mother Mary, they rested under the shade of a tree in Matariya on the outskirts of Cairo; this tree is still preserved today as an important historical site. Again, in Mecca the most important privileges retained by the descendants of the Prophet Mohammed were the guardianship of the well of Zamzam at the Kaaba, and the right to give water to the pilgrims who come to visit the Holy House of God.

Water, in fact, has been regarded as holy by the Arabs, and in view of its supreme importance to their very existence this should not surprise us. For Muslims, water is the means of purification before prayer, while to Christians, water is an indispensable ingredient in baptism. Again not surprisingly, there is a very close association in the Arab mind between water and the sand of the desert: when water is not available for purification before prayer, the clean desert sand may be used instead.

The availability of fresh water has naturally affected the organisation of Arab society and the mobile or settled character of its many groups. Where sources of water are scattered, the nomadic Bedouin Arabs travel in search of it and pitch their tents wherever they find it. The sedentary Arabs are those who have settled by permanent sources of water from river or spring, or where regular seasonal rains occur, and have built houses there and cultivated the land. In nearly every Arab country, though in varying proportions, the historic division of populations still exists, for every cultivated area has a boundary of harsh desert.

★　　　★　　　★

The most suitable areas for the development of agriculture and for the consequent growth of population lie, of course, in alluvial plains. In the Land of the Arabs there are two such types of plain: river flood-plains and coastal strips.

The most notable flood-plains are those of the Tigris and the Euphrates in Iraq; the Orontes in Lebanon and Syria; the Jordan; the Nile; and some smaller rivers in Syria and in the countries known as the Maghreb (the West): Tunisia, Algeria, and Morocco.

The coastal plains include those of the Arabian Peninsula, on the Arabian Gulf,

16

Roman cistern in use today in Libya.

Ancient *noria* wheels in Syria.

the Arabian Sea, and the Red Sea, and the coastal plains of Somalia on the Indian Ocean; but on the shores of the Mediterranean there is an almost continuous coastal strip which runs from northern Syria, through Tyre in Lebanon and Acre in Palestine, across northern Sinai to the Nile Delta, and thence along the coast of North Africa, where it is broken by the Jebel al-Akhdar in Libya and occasionally by promontories of the Atlas in the Maghreb, but widens in places to form broad plains such as those of Benghazi and of Sirte. It leads ultimately to the coastal plain of Morocco, which lies on the Atlantic.

Thus, a natural corridor unites the extreme west of the Arab World to the east. The corridor is not wholly continuous, nor is it always easily traversable, since parts of it are desert, but for many centuries travellers have followed it, including the famous Ibn Batuta, who in the fourteenth century AD made two journeys from Tangiers, his birthplace, along the Mediterranean coast to Syria, Arabia, and Iraq, from where he later set out for China.

The Land of the Arabs may be considered, then, as a geographical entity. In the same way, the Arab people, though by no means ethnically homogeneous, have a unity which has been forged in history by the language, the customs, and the traditions which they have in common.

Scholars have tried to trace the ancestry of the Arabs to the races which inhabited parts of the region in pre-history: to the Assyrians, the Babylonians, the ancient Egyptians of Pharaonic times, and other peoples of antiquity. Such scholarly researches, though they are admitted by scholars to be quite inconclusive, have sometimes been used to serve transient political interests in recent years, during the Arabs' struggle for freedom and independence. Thus, Egyptian nationalism, resisting British imperialism, appealed to the Pharaonic theme; the Babylonian and the Assyrian associations were used in Iraq and Syria against British and French domination; while the French colonisers attempted to divide the peoples of the Maghreb by distinguishing Berbers from Arabs.

Such unrealistic distinctions have been extended to the point of flatly denying the Arabism of certain Arab countries. Such arguments were used particularly against Egypt, when the July Revolution of 1952 began to provide a focus and a bulwark for the Arab nationalism which seemed a threat to some Western interests. If it could be shown that there was no true Arab nation, there could hardly be grounds for Arab nationalism. Yet no similar arguments have been adduced as evidence that there is no American nation, although the enormous variety of races which form the United States have come together into a nation within the last two hundred years. They are all Americans, just as the British are British in spite of their widely mixed and often indeterminate ancestry.

To any unprejudiced observer, it is quite obvious that the Arabs who now inhabit the Land of the Arabs, who speak the Arabic language and who are united by a common history as well as by common aspirations, constitute a single people whose fundamental unity was established many centuries ago. Ibn Khaldun, the great Arab philosopher and the founder of sociology, who was born in Tunis in AD 1332 and died in Cairo in 1406, adopted a two-fold classification of the peoples of the Land of the Arabs, dividing them into Arabs proper (*Aruba*) and the Arabised

(*Mustaaraba*). This was the rough classification that had been devised after the rise of Islam in the seventh century AD and the creation of the great Arab nation in the following few centuries.

The Arabs proper were the original inhabitants of Arabia who emigrated from the Peninsula after the birth of Islam and conquered territories from India to southern France. With the rapid spread of the Islamic religion, however, more areas became Arabised and there was constant intermingling between the Arabs and various other peoples of the vast conquered territory. It is, of course, true that during this period of fusion some peoples held fast to their non-Arab identities. Persia, for example, resisted Arabisation and reverted to her own language, while retaining the Islamic creed. Similarly the Spaniards put up a fierce struggle and eventually forced the Arabs out of Andalusia. But the rest of the huge occupied area came to form what we call today the Land of the Arabs, or the Arab World; and this great land was inhabited by people who were forged into a single entity by the vital force of Arabism. This entity remains, despite the later subdivisions into various states.

The Land of the Arabs had to endure a number of invasions and periods of foreign domination from the eleventh century onwards, first by the Tartars, who were later defeated at Ain Galoot by the Egyptians. Then came the Crusaders, who were eventually repulsed by the combined forces of Syria and Egypt, with Jerusalem recaptured by Saladin. Later still, the Arabs came under the sway of the Ottoman Empire. In much more recent times, Britain, France and Italy ruled over vast tracts of Arab territory.

In spite of invasions and occupations, however, the Arabs were able to maintain their identity, while the Land of the Arabs retained its underlying common culture as well as its natural potential. The various invaders did their best to modify or to destroy the characteristics of the Arabs, always without success.

The French experience in Algeria is a particularly noteworthy case. There, after 130 years of colonisation, during which the Algerian people were brought up to speak French rather than Arabic, we are now witnessing an extraordinary Arab revival which demonstrates that the underlying elements of Arabism are stronger than any outside influences. As soon as she had regained her freedom, Algeria reasserted her Arabism by the use of the Arabic language in education, government, the press, radio, television, the theatre, the cinema and in all other aspects of public and official life. In short, Algeria was immediately reunited spiritually with the rest of the Land of the Arabs, even though for so many years she had been claimed as a province of France.

These examples should amply demonstrate the unity of the Arab lands and the Arab people. They constitute indeed a single organic entity. A popular theme, often expressed by Arab poets during periods of struggle against oppression, was that a cry uttered by an Arab on the banks of the Barada in Syria was bound to be heard by a fellow-Arab on the banks of the Nile. It is to the ancient nation through which that great river flows that we turn first in our detailed study of the Land of the Arabs.

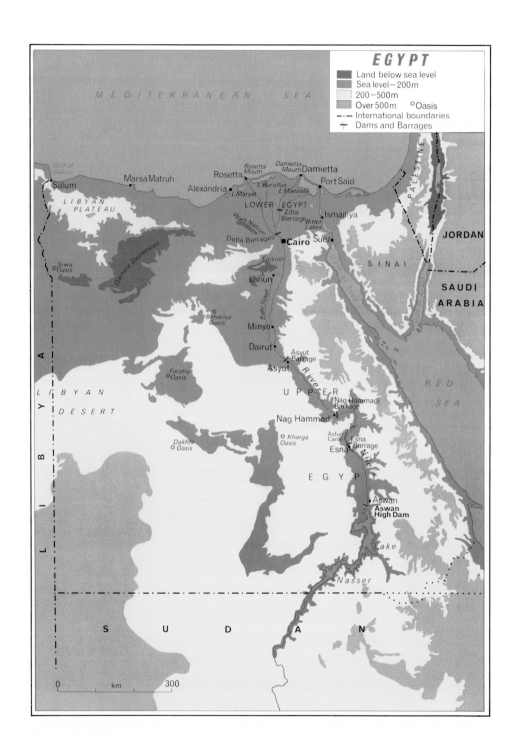

EGYPT

Land below sea level
Sea level – 200m
200 – 500m
Over 500m   ○Oasis
–·–·– International boundaries
Dams and Barrages

MEDITERRANEAN SEA

Gulf of Salum

Salum

Marsa Matruh

Rosetta   Rosetta Mouth   Damietta Mouth   Damietta
Alexandria   Port Said
L. Burullus   L. Manzala
L. Maryut
LOWER EGYPT
Zifta Barrage   Ismailiya
Wadi Natrun   Bitter Lakes
Delta Barrages   Suez
Cairo

LIBYAN PLATEAU

Siwa Oasis

Qattara Depression

Faiyum

Lahun

SINAI

PALESTINE

JORDAN

SAUDI ARABIA

Bahariya Oasis

Minya

Dairut

Bahr Yousef

Asyut Barrage
Asyut

UPPER

RED SEA

Farafra Oasis

LIBYAN DESERT

LIBYA

Nag Hammadi Barrage
Nag Hammadi

Dakhla Oasis

Kharga Oasis

Asfun Canal   Esna Barrage
Esna

EGYPT

River Nile

Aswan
Aswan High Dam

Lake Nasser

SUDAN

0   km   300

The Arab Republic of Egypt

The Arab Republic of Egypt extends from the Mediterranean Sea southwards for over 1,000 kilometres to the border with Sudan, and is almost the size of Spain and France put together. To the east lies the Red Sea; to the west the boundary between Egypt and the Libyan Arab Republic runs in an almost straight line north to south across the Libyan Desert. The area of Egypt is about 1,000,000 square kilometres, mostly uninhabited, while the population of about 34,000,000 occupies only 4 per cent of the total land area, 99 per cent of them living in the Nile Valley and its Delta.

Two physical features dominate Egypt: the huge expanses of desert, and the great River Nile which bisects the country and on which its people depend for their existence. The narrow valley of the Nile, never more than about 15 kilometres wide and often much less in the south of Egypt, opens out into a great delta north of Cairo, and the country is generally thought of as divided into Upper Egypt (the Nile Valley proper) and Lower Egypt (the Nile Delta). On both sides of the river lie vast deserts, with oases in artesian depressions here and there, where water is near to the surface. As in the Sinai Peninsula, where there are no perennial streams, it is these oases which alone enable the scattered nomads and the rare small settlements to survive away from the life-giving Nile. The water, when free from salt, can be used for agriculture, in the western desert particularly. In the large oases of the 'New Valley', where there is fresh water, recently-drilled deep wells will support large agricultural projects with the irrigation of some 17,000 hectares of arable land.

Egypt is one of the most arid countries in the world, with an average rainfall of 15 mm in Upper Egypt and 150 mm in Lower Egypt and the Delta. Summer temperatures are very high indeed, sometimes reaching 49°C in the Western Desert, although on the Mediterranean shore the maximum is a mere 32°C. Winters are warm with cool periods, and snow has occasionally been known to fall. One well-known feature of Egypt's climate is the sand-laden wind known as the *khamsin* (elsewhere the *ghibli*), which can reach speeds of 145 kilometres per hour as it blows northwards across the intense heat of the deserts of Africa during the early summer months.

When the leader of the Islamic Conquest, Amr Ibn el-Aas, first set eyes on the land of Egypt, he described it as 'a green emerald'. Egypt is, as Herodotus said over two thousand years ago, when the country was already steeped in history, 'the gift of the Nile'. Apart from the scattered desert oases, the life of the country is sustained by the 2,500,000 hectares of fertile soil brought down as silt by the river from its sources in Equatorial Africa. Thus we should begin our study of the land of Egypt with a short account of the development of the country and of the use that Egyptians have made of the Nile through their long history.

THE NILE

The sources of the River Nile remained shrouded in mystery for thousands of years. In ancient times it was believed that it flowed from somewhere in the Mountains of the Moon near the Equator. The mystery remained unsolved until the discovery of the sources of the Nile in the nineteenth century—an event which aroused as much popular and scientific interest at the time as the exploration of space does today. News of the various expeditions and explorers—Burton, Speke, Baker and the rest—was eagerly awaited in Egypt as well as in Europe and America. The source of the Nile was in some ways to the Victorians what the philosopher's stone had been to earlier generations, with the great difference that the secret of the Nile could be, and was, unlocked at last.

But for millennia innumerable human beings owed their lives to the Nile, and specifically to the all-important annual Nile flood (see graph on p. 23), without being unduly concerned about its precise place of birth. The Nile flood begins in June; the river continues to rise until the end of September and gradually subsides during October and November until it resumes its normal level. During the flood, the current becomes very swift and the waters take on a brownish hue.

The Nile flood became legendary in the ancient world. The Egyptians of those days gave it an aura of religious sanctity. They celebrated the flood every year with many rituals, notably that of the Nile Maiden. The most beautiful virgin in the land was chosen, bedecked with exquisite clothes and jewellery, and thrown on the festal day into the tumultuous river, the bearer of fertility and life, in gratitude for its yearly flood.

After the Islamic conquest of Egypt, Amir Ibn el-Aas stopped this tradition; but the flood continued to be celebrated without the somewhat doubtful honour accorded to the chosen virgin, though with a dummy figure equally richly adorned.

The ancient Egyptians learned a great deal from the holy river. They became proficient in irrigation engineering, built embankments, barrages and dams, and dug innumerable canals. They even succeeded in diverting the course of the river into Bahr Yusef, thus using its water to irrigate a huge desert area and turning it into one of the most fertile lands in Egypt. This area now constitutes the province of Faiyum. They also became proficient in the planning and engineering of towns and villages, building these on elevated land to avoid inundation by the river during the flood season.

Some authorities believe that the Giza Pyramids were built in order to prevent the desert sands from clogging the river's course. During the *khamsin* season, violent sand-storms blow the desert sand into the Nile and its fertile belt, and the Egyptians had in consequence to dredge the river regularly. The ancient engineers may possibly have hit upon the idea of building the Pyramids on the western bank of the Nile so that the winds would break their force on the sloping sides and deposit their sand harmlessly in the desert around the Pyramids.

No other river in the world was so sanctified, or so commanded the interest of students, as the River Nile. It has inspired writers and poets from Herodotus to Emil Ludwig, and the proverb 'He who drinks from the Nile once must come back to it' is known in many languages other than Arabic.

The River Nile has formed the land of Egypt. There could conceivably be a Nile without Egypt, but no Egypt without the Nile. Besides nourishing its people, the river has also provided them with the means of learning various skills. The ancient Egyptians learned their chronology from the river; they devised a calendar based on the rise and fall of the water and fixed the dates of planting seed and harvesting crops accordingly. This calendar is still strictly observed by Egyptian farmers to this day. Each month in the calendar was associated with a proverb which summed up the prevalent conditions of that particular month. One of the best-known sayings associated with the Coptic calendar, which was derived from that of the Pharaohs, is: 'In the month of Baramhat go to the field and bring in the crop.' (This injunction loses its impact in translation; but in colloquial Egyptian Arabic it rhymes, rather like 'Ne'er cast a clout till May is out.')

Although to modern Egyptians the Nile flood is no longer of great significance, since the flow of the river is now fully controlled by the Aswan High Dam, it must be remembered that until a very few years ago the flood was of vital importance to them. There is a marked pattern in the flow of the Nile, with years of low and high flow as can be seen from the graph below. If the water failed to reach a certain

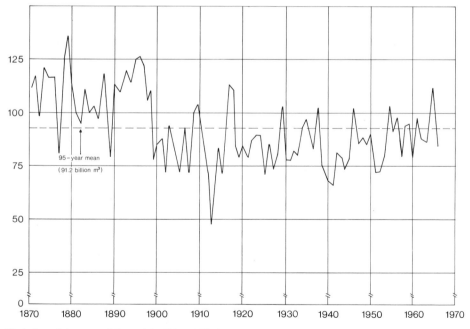

Billion m³

Variation of the natural flow of the River Nile from year to year, 1871–1965. Lack of information has made it impossible to verify the accuracy of the measurements as indicated by the figures (*The Nile Basin*, H. E. Hurst, R. P. Black, and Y. M. Simaika, Cairo: Ministry of Public Works, 1946, 1959, 1966, Vols VII, IX, and X).

The River Nile, Aswan High Dam and Lake Nasser as seen by Skylab astronauts orbiting the earth.

24

level on the Nile Gauge near Cairo (which was built in AD 641), the people were gravely worried, since they then expected, with reason, to go through a year full of hardship until the next flood arrived. There is a complete record of the Nile's flood covering thousands of years. Amin Sami, an Egyptian historian, compiled these data in an interesting book in which he used the figures—indicating the river's rise and fall as recorded by the Nile Gauge—to tell the story of the Nile and of the Egyptian people, year by year.

Similarly, with the aid of these records and statistics and the use of our imagination, we can decipher the main outline of the story of the land and the people of Egypt in relation to the River Nile. In brief, the years of prosperity were accompanied by high floods; and those of misery and famine were years of low flood. The story of Joseph illustrates this: Joseph interpreted the Pharaoh's dream about the seven fat cows and the seven lean ones as indicating that Egypt was to have seven years of prosperity to be followed by seven years of famine. Accordingly, the Egyptians stored the surplus grain during the prosperous years and used it in the years of famine.

Since the Nile is, and always has been, of such supreme importance to Egypt and to its southern neighbour, the Sudan, it would be of interest to give a brief geographical account of the course of the river from its sources to its delta before dealing in detail with the use that has been made of its water by man and with modern Egypt's current and future irrigation plans.

This illustrates the typical seasonal variations of the River Nile before the construction of the High Dam.

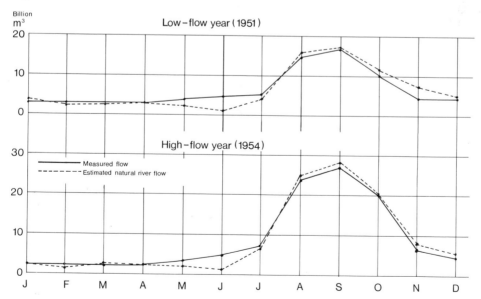

The Nile is 6,700 kilometres long, measuring it from its farthest headstream, the Kagera, which flows from the west into Lake Victoria (1,134 metres above sea-level). Its main basin, extending over more than two and a half million square kilometres and containing more than fifty million inhabitants, can be divided into five major areas.

The Equatorial Lake Plateau, from the main source of the Nile to Lake Albert.

The White Nile (River of the Mountains), until it is joined at Khartoum by its main tributary the Blue Nile, emerges from the north shore of Lake Victoria at Jinja, in Uganda, flowing over the Owen Falls Dam where its outflow is controlled and hydro-electric power is generated. It then spreads itself out into the marshy waters of Lake Kyoga, runs over the Murchison Falls and enters and leaves Lake Albert (670 metres above sea-level) in the north. Another of its sources in Uganda enters Lake Edward and also flows into Lake Albert.

The Murchison Falls with the Nile in full spate.

The Owen Falls Dam at Jinja.

Annual Discharge of the Nile

Years	Period	Mean billion cubic metres	Standard deviation billion cubic metres
30	1870–1899	110·5	17·1
60	1900–1959	84·5	13·5
90	1870–1959	92·6	19·8

Source: Harold Hurst, Robert Black and Yusuf Samaika, *The Nile Basin*, Vol. X, Cairo, Ministry of Public Works, 1965.

The Bahr el-Ghazal Basin

This consists of a vast expanse of almost flat land about 300–450 metres above sea-level, which is almost bisected by the main stream of the Upper White Nile. It extends from a point shortly after the river leaves Lake Albert, still in Uganda, as far as Malakal in the Sudan. Here the Nile flows—or perhaps meanders would be a more appropriate word—very slowly across an alluvial plain and through the monstrous swamps and myriad channels of the papyrus Sudd of Southern Sudan (*sudd* is simply the Arabic word for 'blockage'), where most of its water supply evaporates or seeps into the ground. These marshes and swamps cover an area estimated at 400 kilometres from east to west.

The White Nile Basin

This stretches from Malakal to Khartoum, and the river here runs almost without any inflow from either side after the junction with the River Sobat near Malakal, through the dry central region of Sudan. At Khartoum, the White Nile is joined by the Blue Nile. The Blue Nile originates in Lake Tana in the highlands of Ethiopia, whence it follows a rapid and tortuous course to Khartoum.

Looking upstream towards the Murchison Falls.

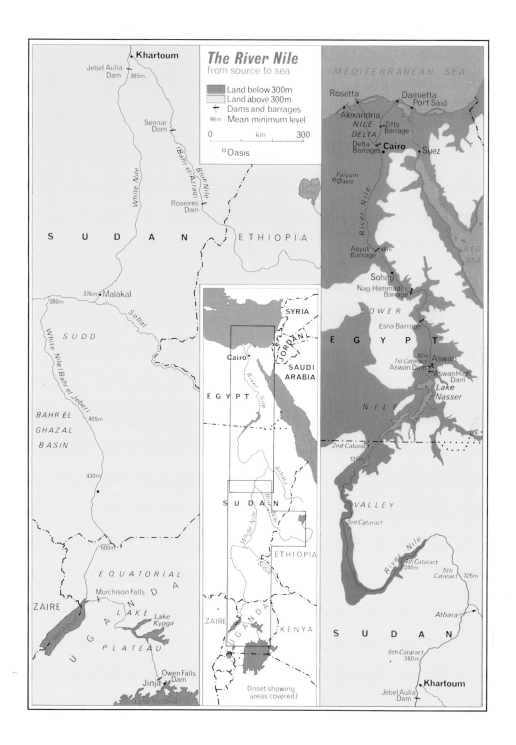

The River Nile
from source to sea

Land below 300m
Land above 300m
Dams and barrages
46m Mean minimum level

0   km   300

○ Oasis

Khartoum

Jebel Aulia
Dam   365m

Sennar
Dam

White Nile

(Bahr el Azraq)

Blue Nile

Roseires
Dam

SUDAN

ETHIOPIA

Lake
Tana

MEDITERRANEAN SEA

Rosetta

Damietta
Port Said

Alexandria

NILE
DELTA

Zifta
Barrage

Delta
Barrages

Cairo

Suez

Faiyum
○ Oasis

River Nile

RED
SEA

Asyut
Barrage   46m

Sohag

Nag Hammadi
Barrage

LOWER

EGYPT

Esna Barrage

1st Cataract   82m Aswan
Aswan Dam

Aswan High
Dam

Lake
Nasser

NILE

Malakal   376m

380m

SUDD

Sobat

White Nile (Bahr el Jebel)

BAHR EL
GHAZAL
BASIN

405m

430m

600m

EQUATORIAL

LAKE

PLATEAU

Murchison Falls

ZAIRE

UGANDA

Lake
Kyoga

Owen Falls
Dam

Jinja

Lake Victoria

SYRIA

JORDAN

Cairo

EGYPT

SAUDI
ARABIA

River Nile

Atbara

Blue Nile

SUDAN

ETHIOPIA

White Nile

Sobat

ZAIRE

UGANDA

KENYA

(Inset showing
areas covered)

2nd Cataract

135m

VALLEY

3rd Cataract

River Nile

4th Cataract
240m

5th
Cataract   326m

Atbara

SUDAN

6th Cataract
360m

Jebel Aulia
Dam

Khartoum

The Lower Nile Valley, from Khartoum to Cairo.

North of Khartoum, the river flows through desert country. One of its sources is the Atbara, which also rises in Ethiopia. The Nile flows in this area over hard outcrops in the Nubian sandstone. Its valley here is narrow, occasionally less than 2 kilometres wide, and the protrusion across the valley of the great crystalline plateau to the east of the river is the cause of the rapids and six famous cataracts that exist in the stretch of river between Khartoum and Aswan. This is the only part of the river where erosion is still continuing, while elsewhere the bed of the river is rising because of the deposition of silt. At Aswan, the great High Dam has created a huge new lake, Lake Nasser, which is also a fish reservoir and stretches far south into the Sudan; while below Aswan the Nile flows in a steep-sided valley, from 16 to 24 kilometres wide, as far as Cairo. Once the Nile enters Egypt there is no addition to its water; losses by evaporation are considerable.

The Nile Delta

The Greeks gave this part of the Nile basin its name, from its resemblance to the capital D in their alphabet (Δ), and the appellation subsequently became a universal geographical term for this phenomenon wherever it occurs. The Nile Delta, some 22,000 square kilometres in area, has been formed through the ages by deposition of sediment by the Nile; and it is, as we shall see later, extremely fertile. Here the Nile splits into two main distributaries, the Rosetta and the Damietta, and innumerable minor streams.

Before leaving this very brief account of the course of the world's greatest river, we should perhaps summarise the Nile's peculiar régime. If the Nile depended only on the water derived from the lake plateau and southern Sudan, there would never have been any great civilisations in Egypt, ancient or modern. The Nile flood would not have taken place, for the simple reason that the river would never have contained enough water to cause it. There is, it is true, a regular flow in the mountainous upper course of the White Nile, thanks to the steady discharge from lakes Victoria and Albert of approximately 26 thousand million cubic metres of water each year. But half the volume of the river is lost to it in the Sudd of Southern Sudan through evaporation and seepage. The River Sobat helps to make up some of the White Nile's losses, but even its flow from the Ethiopian mountains is markedly diminished by swamps.

It is in fact the Blue Nile which saves the situation as it joins its bigger—or, rather, its longer—brother at Khartoum. Its contribution to the annual discharge of the river below Khartoum is twice that of the White Nile and without it there could have been no Nile flood in Egypt. Following the March and April rains in the Ethiopian mountains, the level of the Blue Nile rises swiftly and dramatically (up to 12 metres) between June and September and, aided by the smaller Atbara River (which contributes 22 per cent of the total high water flow), it is the Blue Nile that plays the chief part in determining the annual cycle of river-flow below Khartoum. Below Atbara very little water is added to the Nile. (See graph on page 32.)

Having paid tribute to the Blue Nile for its unique and necessary contribution to the Nile flood-waters, one must note that during the low-water period the White Nile provides over 80 per cent of the river's flow. Without the Blue Nile there would be no Nile flood: without the White Nile, there would be not much more than a trickle in the dry season.

RESERVOIRS ON THE NILE

Napoleon, who occupied Egypt in 1798, conceived the idea of constructing a barrage at the apex of the Nile Delta so as to bring irrigation to that potentially fertile region. His memoirs, written during his stay in Egypt, showed that he foresaw the beginning of the modern system of regulating the Nile waters. But it was Mohammed Ali, the then ruler, who in 1835 put in hand work on the two branches of the Nile. However, the structures (known as the Mohammed Ali barrages) were not a success because of continual maintenance problems, and they were not capable of dealing with the increased amount of water available after the heightening of the Aswan and Jebel Aulia barrages. So new structures, the Damietta and Rosetta barrages, were constructed and completed in 1940. These, sited on the Damietta and the Rosetta branches of the river at the head of the Delta, allow the full discharge of the river to flow equally through its eastward and its westward courses throughout the year, and raise the level of the water upstream so that when the general river-level is low, water can still flow into the canals which irrigate the Delta. In fact, the construction of the 450 and 530 metre-long barrages, on the Damietta and the Rosetta branches respectively, constituted one of the most important water regulation schemes undertaken so far, since they allowed for the full exploitation of the Delta land by means of organised irrigation.

A barrage is a structure of masonry and/or earth built across a waterway, such as a river or tidal estuary, so as to control or regulate the flow of water. A barrage is designed in such a way as to pass through itself the maximum flow of water. Gates, which are nowadays of the radial type, are incorporated into the structure and regulate the level of water upstream, diverting some or all of it into secondary channels. A barrage can act as a dam, in that it can completely close or divert a river. And like a dam it can be used to create a storage reservoir. But unlike a dam a barrage has gates along its entire length, and these gates, when fully open, allow the water to flow normally in its original channel. A weir is a smaller version of a dam. Most barrages have fish-ladders, and many have navigation locks.

At the end of the last century, work was started on another major scheme: the construction of the Aswan Dam, which began in 1898 and was completed in 1903 on a site where the first cataract is formed. In 1908 work began on the first raising of the height of the dam, which was finished in 1912. After its second raising, which was completed in 1934, the dam, now about 2 kilometres long, reached a storage capacity of 5·5 billion cubic metres of water and a depth when filled of 120 metres.

The construction of the Delta Barrage and the Aswan Dam went hand in hand with several auxiliary projects, including the following:

The Zifta Barrage on the Damietta branch of the Nile, between the towns of Zifta and Mit-Ghamr (1902).

The Asyut Barrage, which began and ended at the same dates as the Aswan

Dam (1898–1902). From this barrage leads the Ibrahimiya Canal, 400 kilometres long and the largest in Egypt. The barrage itself has 111 sluice gates,
and was remodelled (barrage and regulator) between 1935 and 1938.

The Esna Barrage (1906–09).

The Nag Hammadi Barrage (1927–30).

All these schemes were of fundamental importance to the regulation of irrigation
in Egypt. But work extended also to the Sudan, where in 1925 the Sennar Dam was
constructed in granite, 3,000 metres long, across the Blue Nile. This was followed
by the construction of the Jebel Aulia Dam, also in the Sudan, which was completed in 1937, with an overall length of 5,000 metres. The purpose of the latter
was to provide a reservoir which could replenish the Aswan Reservoir when the
water there fell below a predetermined level.

Each river's monthly contribution to the floods of the Nile (after J. Oliver *et al*).

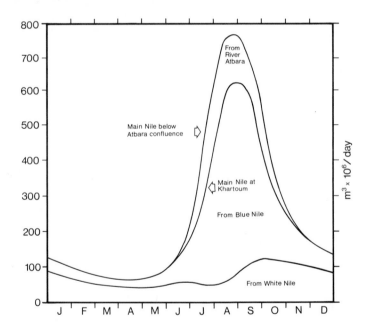

The start of the construction of the Aswan Dam.

The Aswan Dam nearing completion.

The Nag Hammadi Barrage.

However, despite this century of constructive effort, the Nile waters were still not fully harnessed to irrigate the arable lands of Egypt, nor to impound water during the flood period for use when the river was low. From 1936 to 1952 all further work on the regulation of the Nile water ceased. Indeed, the construction of the Jebel Aulia Dam itself served only to put greater strain on the Aswan Dam, which nobody dared to raise for a third time lest it should collapse. So man's great works on the Nile remained incomplete until the High Dam was built to the south of Aswan.

34

One of the two original Nile Delta Barrages, which were replaced by new structures in the late 1930s.

The dramatic increase in the population has emphasised the need for maximum use of land, as shown in the graph.

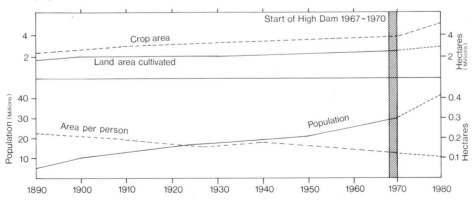

An unlined irrigation canal in Upper Egypt, where the ribbon of cultivation is extensive.

A *felucca* with cargo on the Nile near Aswan. Navigation locks on dams and barrages give access to vessels.

Now, for the first time in the history of the land of Egypt, irrigation work is being fully controlled. The entire area of arable land is being put under the perennial irrigation system, and the days when thousands of acres of land in Upper Egypt were submerged during the flood seasons have gone for ever. Instead, the flood water is being diverted to new courses in the heart of the desert to create new arable areas in Tahrir District and alongside the desert road between Cairo and Alexandria.

The story of the building of the High Dam will go down as one of the most exciting episodes in the political history of the modern world. Its construction gave rise to grave political crises which led to the 1956 Suez War; but the dam was built in spite of all the difficulties that stood in the way. Now it stands as one of the most important construction works undertaken in the twentieth century. It is considered to be the second largest dam in the world as far as capacity of its reservoir is concerned, and ranks eighth largest in generation of electric power. The tremendous impact on the political and economic life of the country which this dam has created will be recorded as a great achievement in the history of the Egyptian nation, alongside the 'stupendous monuments' erected by the pharaohs.

The Egyptian people finished this, the last of the great barrages and dams on the Nile, on 15 May 1970. The river's course was diverted to bring about a significant change in the very nature of the Egyptian land. Basin-lands were converted to perennial irrigation; that is, from half to full productivity, while thousands of acres of yellow sand, extricated from the clutches of the desert, are now covered with wheat or maize which flutters in the breeze of prosperity brought about by the High Dam.

The High Dam was built 6 kilometres upstream of the old Aswan Dam, and 640 kilometres south of Cairo. It is 3,600 metres long with a roadway 40 metres wide at the crest, and it is 550 metres from bank to bank at the base. It has a storage capacity of 157 billion cubic metres, thus providing a constant supply of water which fully satisfies present requirements and allows for extensive future expansion of the work of reclaiming desert areas. With the elimination of the flood season now that the High Dam has been completed, the old irrigation canals are being replaced by new gravity flow systems.

Moreover, the electric power generated from the High Dam, which is already being transmitted, will change the face of Egyptian rural life, quite apart from playing an essential role in the industrialisation of agriculture. Already, High Dam electricity is finding its way to Egyptian villages which hitherto enjoyed no power of any kind other than that derived from the sweat of man and beast.

The dam is also forming an enormous artificial lake, Lake Nasser, extending 500 kilometres southwards across to the Sudan, 27 kilometres wide at its widest and 3 kilometres at its narrowest. When full at the sluice gates the depth is $91\frac{1}{2}$ metres. This reservoir has become the centre of an expanding fishing industry which, given time, may replace the sardine catch that has been one of the 'casualties' in the Mediterranean Sea because of the dam.

The effect of the High Dam upon the development of Egyptian society is incalculable. As to its effect upon the land itself, suffice it to say that the High

Dam is the corner-stone of all the agricultural projects contemplated for the fore-seeable future.

As with all projects of this magnitude, some words should be added here regarding the problems usually associated with irrigation dams of large reservoir capacity: evaporation, which is great in arid regions, seepage to neighbouring areas, sedimentation of silt, and degradation or erosion. Water stored above the High Dam loses nearly all of the material suspended in it; the increased velocity of pure water below the dam begins the process of erosion of the alluvial bed to make up for the loss of material deposited in the reservoir. This in time reduces the level of the water, with an increase in the pressure on the barrages north of the High Dam to the mouth of the Nile.

This scouring action of the Nile, which will in future drop its load of silt in Lake Nasser and tend to pick up silt from its channel below the dam, could, according to Marion Clawson *et al.*, 'create new serious problems for the downstream dams or barrages'. Only time will tell what the changes to the regimen will entail.

Impounded Nile waters being released through the partially completed dam.

Foundations.

The hydro-electric machinery.

Part of the high-tension distribution system.

The New High Aswan Dam. Tipping lorries bring rock rubble.

Rock tunnels in the Dam.

Steel reinforcing and pre-cast concrete tunnelling.

Concrete construction work on the Dam.

Controlling the flow of water through the completed dam.

IRRIGATION AND DRAINAGE

Irrigation

There are two main types of cultivated area in Egypt: first, land under perennial irrigation which produces more than one crop a year; and second, basin-irrigation lands. Crop rotation was adapted to this form of irrigation, but it meant that only one crop could be cultivated in the winter, as the Nile is at its lowest level in the summer. Basin irrigation is the ancient system whereby the river waters are led off into specially-prepared basins separated from one another by banks of earth. Perennial irrigation depends on the use of barrages and dams, and on permanent water storage.

The area of perennial irrigation has been increasing steadily at the expense of basin irrigation as a direct result of the construction of dams, barrages and other schemes. In 1866, the area of basin-irrigation land was about 800,000 hectares as compared with 1,200,000 hectares of perennial irrigation lands. In 1922, however, the area of perennial irrigation had increased to 1,600,000 hectares while the area of basin lands decreased to 640,000 hectares. This trend has continued over the years, and by 1954 the area of basin lands had fallen to 280,000 hectares while the area of perennial irrigation land had increased to 2,600,000 hectares. After the construction of the High Dam, however, the basin-irrigation system is being abandoned altogether, and all cultivable land is being converted to perennial irrigation.

Since the dawn of its history, Egypt has used the Nile water for irrigation. The

A diesel pump raises the water for immediate use on the fields.

Basin irrigation, where stagnant flood water is likely to induce soil salinity through the accumulation of sodium and magnesium chloride. This system is now almost a method of the past.

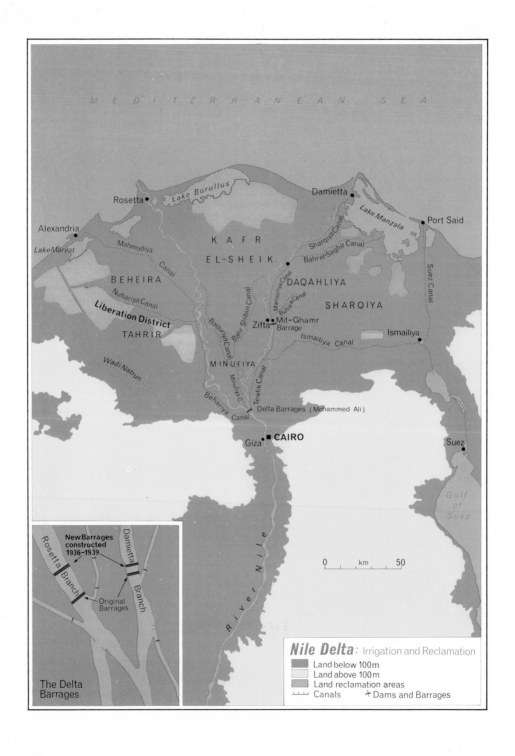

MEDITERRANEAN SEA

Rosetta

Lake Burullus

Damietta

Lake Manzala

Port Said

Alexandria

Lake Maryat

Mahmudiya Canal

KAFR

EL-SHEIK

Sharqiya Canal

Bahr el-Saghir Canal

Suez Canal

BEHEIRA

Nubariya Canal

Liberation District

TAHRIR

Wadi Natrun

Baguriya Canal

Bahr Shibin Canal

Mansuriya Canal

Buhiya Canal

Zifta

Mit-Ghamr
Barrage

DAQAHLIYA

SHARQIYA

Ismailiya

MINUFIYA

Minufiya C.

Tewfik Canal

Ismailiya Canal

Behariya
Canal

Delta Barrages (Mohammed Ali)

Giza ■ CAIRO

Suez

River Nile

Gulf
of
Suez

0   km   50

**The Delta
Barrages**

New Barrages
constructed
1936-1939

Rosetta Branch

Damietta Branch

Original
Barrages

Nile Delta: Irrigation and Reclamation

Land below 100m
Land above 100m
Land reclamation areas
Canals   Dams and Barrages

basin system was predominant throughout Egypt until towards the middle of the nineteenth century. When the Nile rose, the water was let into the basins to flood the land to an average depth of $1\frac{1}{2}$ metres; and the water was held there for forty-five days on average (maximum sixty days). After the river had fallen sufficiently, the water was drained back into it by gravity flow; but the land would by then have been saturated in preparation for a long period of thirst, and the silt carried by the flood water from the Ethiopian mountains would have given the soil added fertility and compensated it for what it had lost in the previous year. Alluvial soils are formed from materials deposited by running water in stream valleys or deltas. The soils of the Nile Valley and Delta are examples of this type.

The lands of Upper Egypt, with the exception of Faiyum, were divided into basins separated by longitudinal earth embankments or levées extending from the bank of the river to the edge of the desert. The basins or compartments were fed by canals which drew water from the Nile in the flood season; but when the dry season came the water dropped below the level of the canal intakes, leaving them high and dry. There were no barrages to regulate the flow of water from the river into the canals, with the exception of the improvised earth and stone dams on the intakes, which were removed every year to let the flood water into the basins. This was usually done in the second week of August after the harvest of the wheat, barley and fodder crops (best suited to this type of irrigation). Each group of basins was fed by a canal; these varied considerably in size, from very small ones supplying one or two basins only to large and vastly important canals serving wide areas. After one and a half months, on average, the water from each group of basins was drained back into the river through a special outlet, or drainage canal. The draining took place at about the beginning of October, becoming a little later farther north along the river's course.

That system of irrigation was successful in Egypt because the changing conditions of the river coincided neatly with the weather in the country. That is to say, the Nile flood arrived in the season when the weather was most favourable for the cultivation of most crops. The topography of the Nile Valley, too, is favourable to the basin-irrigation system, since it slopes down to the east and west from its highest point on the banks of the Nile; thus allowing the water to run down easily to the extreme ends on both sides of the river and making simple the eventual drainage of the basins through the special escape-routes already mentioned. The river flows closest to the eastern valley wall; the cultivated area is therefore greatest west of the Nile.

The system was not confined to Upper Egypt but was also used in the southern parts of the Delta. The northern Delta is flat and in places lies below sea-level, causing lakes and marshes to form and getting very close to the underground water-table. In such places the soil is highly saline and barren. Among these wide expanses, known as *barari*, the only habitable areas that remained were the banks of the Rosetta and Damietta branches of the Nile, and of other minor branches, which formed oblong depressions in the midst of these desolate expanses.

During the centuries in which the basin-irrigation system was predominant in most parts of Egypt, the country bordering the Nile resembled, in the flood season, a huge lake in the heart of the desert, submerged to an average depth of $1\frac{1}{2}$ metres and divided up by thick walls of earth. The villages stood amidst the water like islands. This accounts for the fact that the Egyptian village has always been compactly built, with the farmers' houses separated only by very narrow alleys. Boats were the only means of transport between the villages during the flood season. It was a season of comparative idleness and rest imposed by nature upon the peasants; but as soon as the land dried up they resumed their toil. They cultivated only one crop every year, usually cereals, except in those limited areas very close to the banks of the Nile which could be easily irrigated by water lifted from shallow wells by the *sakia* (water-wheel) or the *shadouf* (a bucket on the end of a long lever).

Cultivated land under controlled irrigation.

It was indeed this that gave the Delta, in those days, such an advantage over Upper Egypt. For in the Delta, the land was sown with summer crops which were irrigated by water lifted either from the canals or directly from the branches of the Nile where the water-table in the dry season was nearer to the surface than it was in Upper Egypt. Land so cultivated with summer crops had, of course, to be protected against the flood water by the familiar earthen embankments which the farmers built around their fields.

Towards the end of the rule of the Mamelukes (AD 1250-1517), the agricultural situation in Egypt had deteriorated to a great extent. Irrigation canals were left to be filled in by accumulating silt, their banks were washed away by the flood water which inundated the adjacent lands and there was almost no control of irrigation. Water reached some areas in abundance while others were left completely dry.

The Mameluke rulers persisted in their negligence for so many years that enormous tracts of good productive land became sterile and barren. This was the situation found by the technicians who accompanied Napoleon on his Egyptian campaign. They decided that the irrigation system needed a sweeping reform. Egypt has indeed benefited from the projects and plans which they devised, and which were later carried out by Mohammed Ali to try to increase the area of perennial irrigation. These technicians thus laid the foundation of a new water policy, with the accent on digging irrigation canals, boring for water (artesian wells), installation of pumps, and constructing dams and barrages.

The Delta was the field of their first experiment. They began the work of converting the Delta lands from basin to perennial irrigation by digging new canals and deepening the existing ones so that water would continue to flow in them even during the summer. The most outstanding effort in this direction, under Mohammed Ali, was the digging of the Mahmudiya Canal between the Nile and Alexandria, which was officially inaugurated in January 1820.

This was a very successful undertaking, not only reviving Alexandria as a port, but increasing the area of land irrigated by this canal from 1,600 hectares to the present figure of more than 120,000, that is about 5 per cent of the total cultivated land in Egypt. Moreover, the construction of the great Nile Barrage at the apex of the Delta, which was completed in 1861, was accompanied by the digging of three major irrigation canals: the Behariya Canal for irrigating the western sector of the Delta, the Minufiya Canal for the central sector, and the Tewfik Canal for the area east of the Damietta branch of the Nile. After the construction of the Delta Barrage, the difficulties of lifting the water from the irrigation canals to the higher landsurfaces during the dry season were overcome, since the very purpose of the barrage is to raise the level upstream, so that when the river is low water can still flow into canals provided that they are dug to a reasonable depth and properly maintained.

In fact, as we have already seen, the Delta Barrage was the first major construction along the whole course of the Nile to be built with the object of controlling the discharge of the river and of providing irrigation water for summer cultivation. Several other projects that served the same purpose accompanied and followed its establishment. In 1860, the Ismailiya Canal was dug to serve as a navigable waterway connecting the Nile with the Red Sea, and to provide drinking water for the

47

A *shadouf* method of irrigation using a bucket and lever to raise water from the level of the canal to the field. This involves standing in the water for long periods of time, thus increasing the danger of contracting bilharzia from the liver-fluke.

new settlements along the projected Suez Canal, which was opened in 1869. The Ismailiya Canal branches off the Nile at Benha just north of Cairo and follows the edge of the desert to Wadi Tumeilat, where it takes an eastward course to the town of Ismailiya. Just before it reaches Ismailiya it splits into two branches, one running north to Port Said and the second south to Suez.

The Ibrahimiya Canal, perhaps the longest man-made irrigation canal in the world, was dug in 1873. It branches off the main course of the Nile at Asyut and cuts through the higher land near the river bank for a distance of 267 kilometres. It has an intake varying between twenty and seventy million cubic metres a day, compared with an average of nine million for the River Thames at Teddington. It was intended at first to irrigate the sugar-cane plantations of the Khedive of Egypt; but it eventually provided irrigation water for the summer cultivation of 235,000 hectares as well as for the basin irrigation of 170,000 hectares—a total of about 400,000 hectares, or 17 per cent of the total cultivated area in Egypt. In 1873, it was connected at Dairut with the Bahr Yusef Canal, which runs through the provinces of Asyut, Minya and Beni Suef before it leaves the Nile Valley at Lahun and turns westwards to the Faiyum depression.

It was obvious, however, that the river's discharge was not enough to meet the requirements of total conversion to perennial irrigation. Efforts were therefore made to find methods of storing the river's water in order to achieve the dual purpose of increasing the arable area and converting it to perennial irrigation. It was to serve this purpose that the Aswan Dam was built. Its completion in 1902 coincided with the construction of the Asyut Barrage, which regulated the flow of water into the Ibrahimiya Canal, thus making possible the perennial cultivation of 400,000 hectares and the basin cultivation of a further 160,000 hectares. In 1937 the Asyut Barrage was remodelled to meet the increasing need for irrigation water in central Egypt.

Headworks on the Ibrahimiya Canal.

That same year saw the completion of the Zifta Barrage on the Damietta branch of the Nile. This barrage feeds the Bahr Shibin Canal via the Abbasi Canal, and the Tewfik Canal via the canal of Mansuriya.

The Esna Barrage was built in 1908, 160 kilometres north of the Aswan Dam, in order to improve the basin irrigation of land in the province of Qena in the years of low flood, by means of the canals of Asfun and Kelabiya. But even this did not solve the problem of improving the economically valuable lands north of Nag Hammadi. These areas frequently suffered from severe drought in low-flood years. In the space of half a century, from 1863 to 1913, this region suffered from ten years of low flood. The situation was not improved by the Aswan Dam, nor by the barrages of Asyut or Esna, and it became necessary to build the Nag Hammadi Barrage with two head regulators leading to irrigation canals, which reached completion in 1930 (see map, p. 29).

However, the full agricultural use of these areas depended upon increasing the quantity of water stored above the Aswan Dam after its second raising in 1933, and also upon the construction of the Jebel Aulia Dam on the White Nile in the Sudan. Meanwhile, two canals were dug, leading off from the Nile above the barrage to bring water to the irrigation area. These were the Gharbiya (Faroukia) Canal, which was completed in 1930, and the Sharqiya (Fouadia) Canal, opened in 1932. The latter had to be driven through the Ahaiwa Tunnel which was dug in the rocks overlooking the Nile to the north of Nag Hammadi. The Nag Hammadi Barrage thus made it possible to convert 56,000 hectares in the Akhmim, Badari and Abnub districts to perennial irrigation via the Sharqiya Canal; and 120,000 hectares in the provinces of Gharbiya and Asyut, via the Gharbiya Canal.

But in spite of all this it was evident that still greater quantities of stored water were needed for agricultural expansion in Egypt. It was this need that gave birth to the idea of building the Jebel Aulia Dam some 45 kilometres south of Khartoum. The river-bed at this spot is shallow but very wide, so that a great deal of water is lost by evaporation. The project was originally proposed in 1920 with the object of increasing the arable area in Egypt to two million hectares by 1935 (assuming the completion of the dam by 1925), to two and a half million hectares by 1945 and to almost three million by 1955. The last figure represented what was estimated to be the maximum possible expansion in the Nile Valley and the Delta regions of Egypt.

The original project was revised later to try to reduce its cost; but it was not until 1932, when it was finally approved, that work actually began on its construction. As already stated, it was completed in 1937 and filled to capacity (375 metres) by 1942, and it is still, at $6\frac{1}{2}$ kilometres, the longest dam in the world. The Jebel Aulia Reservoir is drawn upon in February and it is usually empty by May.

Every year from 1885, the Egyptian Department of Irrigation used to construct an earthen bank (or *sudd*) across the Rosetta branch of the Nile at Idfina. This was done at the end of the flood season, when the flow of the river was shut off at the Delta Barrage. The bank prevented saline efflorescence from running inland, and it also enabled seepage and drainage to be gathered back into the river and used for irrigating the northern parts of the province of Kafr el-Sheik. The quantity of water collected in this manner is estimated at 250 million cubic metres. In order to avoid the yearly repetition of this primitive exercise, however, the bank has now been replaced by a masonry barrage at Idfina on which work was finished by the end of 1951.

Salt efflorescences are common in desert soils: that is, the surface hardens temporarily into a light crust that will allow a tractor over what otherwise would be loose sand.

It may be useful here briefly to summarise the main arteries of Egypt's irrigation network.

The Western Delta

The area of arable land is about 260,000 hectares. It is served mainly by the Mahmudiya Canal and the Behariya Canal with outlets from the Nile above the Delta barrage. (The Behariya is also called the Khatabba and the Sahel Morkos before it joins the Mahmudiya Canal.) It feeds the Nubariya and the el-Hager canals which branch off it, as well as the Abu Diab. All these water the southern part of the province of Beheira. The Behariya Canal feeds the eastern and western Khandak canals which water the central part of the same province.

The Central Delta

The cultivated area is about 730,000 hectares. This large tract is fed mainly by the Minufiya Canal, with its outlet from the Nile at the head of the Delta, running for about 25 kilometres before it branches off into the Bahr Shibin and el-Baguriya canals. The former is the more important of the two, since it feeds a large network of smaller canals which water the greater part of the Central Delta. The Baguriya Canal, known in its northern part as the Gadahiya, waters the West-Central Delta.

Before it branches off into the Bahr Shibin and the Baguriya canals, however, the Minufiya Canal feeds the Nenaiya, which runs parallel to the Rosetta branch of the Nile as far as Kafr el-Zayat, and the Sersuriya Canal, which ends just north of Shibin el-Kom.

The Eastern Delta

The total cultivated area is about 1·2 million acres. This region is irrigated by the Ismailiya, Sharqiya and Buhiya canals with outlets directly from the Nile; but the most important artery feeding the area is the Tewfik Canal, with its outlet north of the Delta Barrage. This canal was dug with the primary object of feeding the Bahr Mones Canal in the province of Sharqiya, as well as the Buhiya, el-Bahr and el-Saghir canals in the province of Daqahliya. At Mit-Ghamr, the Tewfik assumes the name of Mansuriya until it reaches the town of Mansura, where it branches off into the Bahr el-Saghir, which ultimately flows into Lake Manzala, and to the Sharqiya, which runs parallel to the Damietta branch of the Nile.

By far the greatest source of Egypt's water is, of course, the Nile, and the control of supplies is essential. The High Dam now provides a regular supply, but above this point, in Sudan and in other African states, wastage is enormous: of the rainfall which drains into Lake Victoria, only 8 per cent reaches Egypt, the rest being lost or absorbed by evaporation in marshland and swamp. The Nile Waters Agreement between Egypt and Sudan provides for work to reduce these losses and to share the resultant increase in supply between the two countries. It is hoped that of an estimated annual loss of 36 billion cubic metres, a half may be saved; thus, to Egypt's annual flow of 55·5 billion cubic metres, another 9 billion may be added, largely by control of loss from the regions of Algabara, Alzaraph, and Bahr el-Gazal, and from the basins of the Soubat and the White Nile.

The second source is subterranean water, of which the amount, still unknown, is certainly vast. Research must be undertaken to determine the hydrological and geological characteristics of the land in Upper Egypt and in the Delta, the probable quantities of water available, and the effect on it of evaporation and of the High Dam. Already, it has been ascertained that beneath the Delta lie some 500 billion cubic metres of drinkable water, and that there is a constant movement of subterranean water from north to south.

Two lesser sources must be mentioned: rain, and the desalination of sea-water.

The averages of rainfall to which we referred earlier, analysed in greater detail, show that the Mediterranean coast to the west of the Delta may have up to 200 mm per year, the coast east of the Delta and the middle Delta have between 25 and 50 mm, and parts of Upper Egypt receive no rain at all. Since established criteria suggest that rainfall less than 700 mm per year cannot nourish agriculture, the most that can be expected from rain in Egypt is the provision of some pasture.

Experiments with desalination of sea-water have begun in Egypt, after their success in other countries. At present some drinking-water can be provided, but until scientific work has been much extended there will be little promise of the economic use of this source for irrigation.

Drainage

The main object of Egypt's water policy, that of providing perennial irrigation, was in fact realised to a great extent, though it gradually became apparent that the improvement of irrigation facilities was being achieved at a high cost. More frequent irrigation raised the underground water-table with consequent deterioration of the soil. It was necessary therefore to draw up a consistent drainage policy to go hand in hand with the irrigation projects. Such a policy was decided upon somewhat later in the day, for until 1885 drainage canals in Egypt were both few and inadequate. Since then, however, this state of affairs is being gradually rectified by the use of buried tile or plastic drains which will lead to greatly improved soil-water relationships. This in turn can increase crop yields by up to 50 per cent. And if buried drains ultimately replace all the old surface drains, then enough land (somewhere in the region of 12 per cent) can be saved to pay for the project.

From the point of view of drainage, the arable land in Egypt can be divided into the following areas:

Upper Egypt

Land in Upper Egypt which is still under basin irrigation does not need artificial draining. This is because the river itself acts as a large drainage canal for the narrow strip of arable land available along its banks. But provision must be made for draining the land that has been converted to perennial irrigation in order to conserve the fertility of the soil and to safeguard adjacent lands against seepage.

Central Egypt

The level of the land here is higher than the bed of the Nile, and the cultivable area forms a narrow strip along the river-banks. For these reasons no provision was made for drainage when this land was converted to perennial irrigation. After years of continuous irrigation, however, the need for drainage became apparent. Some drainage canals were dug where it was feasible and several pumping-stations were established for drainage in places where canal-draining is not practicable.

Faiyum

The land in the region of Faiyum has an adequate network of drainage canals. Drainage by canals is made easy owing to the general downward slope of the land in the direction of Lake Qarun. The only exception to this is an area of about 16,000 hectares to the south-west of the Faiyum Depression where drainage is carried out by pumping.

The Northern Delta Lands

The land in the north of the Delta slopes gradually towards the sea and the inland lakes so that the northern ends of the canals there are below sea-level and must therefore be discharged by pumping. There are eighteen pumping-stations in the north of the Delta to serve this purpose.

The Southern Delta

The land here is at an elevation of 6 metres above the river-level, thus allowing for the natural drainage of the land. But continuous irrigation has caused the underground water-table to rise and it is now becoming necessary to make provision for draining this land to restore its fertility.

Distribution works on a feeder canal.

A sluice gate controlling the dispersal of irrigation water.

THE LAND

In a country such as Egypt, with three thousand years of known history (thirty Pharaonic dynasties succeeded one another), it is particularly difficult to trace back the story of the land and to define with precision the size of the cultivated areas at different periods. So far we have naturally concentrated on the Nile. There were, however, other scattered areas of this ancient country that have been under cultivation in past ages, notably in the Western Desert and some parts of Sinai. The Eastern Desert has never been cultivated since the dawn of history and it is hard to see how it ever could be economically exploited in this way.

During certain periods of Egypt's history, the cultivated parts of the Western Desert were neglected and their population withdrew into the Nile Valley. There were also times when the lands in the Nile Valley and in the Delta, too, were neglected, with the result that irrigation canals and ditches dried up, embankments were destroyed and the population dwindled to a few millions. Indeed, when Napoleon invaded Egypt in 1798, the French scientists with him estimated the population then to be no more than four million. Nevertheless the history of Egypt in modern terms began with Bonaparte.

Yet it was Mohammed Ali's rule (after the tyranny of the Mamelukes) which had lasting results in Egypt. Aided by technical assistance from Europe and the proceeds of exports of cotton, which he was the first to develop, he used these resources for such purposes as irrigation works, the great Delta Barrage (mentioned on p. 31), and its network of canals; the Mahmudiya Canal between the Nile and Alexandria being but one notable example.

Al-Maqrizi, a well-known Egyptian historian, has written a book entirely devoted to the famines and other catastrophes which occurred in Egypt owing to neglect of the land and of flood control. Sometimes there were droughts leading to starvation, while at other times floods inundated the land with equally disastrous results.

The people of Egypt flourished when irrigation was controlled and organised, when canals were dug and embankments were constructed and maintained; but it was the wars and feuds among the Mameluke Sultans which brought about the fatal neglect described by al-Maqrizi in his book. During such periods of unrest, the rulers naturally neglected the lands of the Western Desert: both those in the north-west from Alexandria to Marsa Matruh, and the oases scattered in the heart of the desert—those of Siwa, Dakhla and Kharga. As a consequence many of the old Roman wells which had once irrigated vast areas were filled in. Egypt was once known as the granary of Rome, and it is thought that much of the grain may have come from those western oases; some historians have ventured estimates of their total area at three and a quarter million hectares. The Middle East is the ancient home of wheat and barley; originally it grew wild, being subsequently developed as a domesticated, cultivated crop.

In spite of the increasing interest shown by the State in irrigation control since the beginning of the nineteenth century, from the construction of the Delta Barrage (the Mohammed Ali Barrage) up to the building of the Aswan Dam, it failed to pay adequate attention to the lands of the Western Desert, nor did it make the slightest attempt to reclaim desert areas.

In fact, the interest shown in the cultivation of the land within the Nile Valley itself was half-hearted. It took more than a century, as we have seen, to implement some modern schemes of irrigation. This was not due to any technical or material lack. For years Egypt had numbered among her sons many irrigation engineers of genius. As to material resources, the cost of the celebrations organised by Khedive Ismail for the opening of the Suez Canal was more than enough to build the Aswan Dam, the blue-prints of which were already drawn up at the time. Indeed, several missions were sent to examine the site and to conduct the necessary studies for the implementation of the project. All that emerged were plans for building the famous Cataract Hotel, which was in fact constructed—but not so the dam. Nor did the plans for generating electricity from the cataracts materialise, whereas the Pyramids of Giza and the whole area around them were flood-lit with electric light when they were visited by Khedive Ismail accompanying the Empress Eugenie.

Since the Revolution of 23 July 1952, the whole picture has changed. From that time every resource has been mobilised to ensure the maximum development of the land for the prosperity of all the people. During the last twenty-three years important chapters have been added to the story of the land in Egypt; particularly the building of the Aswan High Dam which ensures full control of the Nile water for the first time in the history of Egypt.

LAND RECLAMATION

In addition to the provision of perennial irrigation and adequate drainage for land already under cultivation, the expansion of the cultivated area was one of the most important aims which Egypt strove to realise after the Revolution. The first of the great schemes directed towards the counter-invasion of the desert was the Tahrir District project.

Tahrir District

Tahrir (Liberation) District lies between the Rosetta branch of the Nile and Wadi Natrun, to the south of the province of Beheira. The Nubariya Canal runs through its reclaimed land at the northern end of the district.

It is divided into three sectors—the southern, the northern and the el-Tahaddi ('challenge') sector. Already a total area of 66,000 hectares has been reclaimed. It will perhaps be interesting as an example to look in more detail at what has happened and is still happening in one of these sectors, the southern.

The reclaimed area in the southern sector amounts to some 21,425 hectares, divided into eleven farms as follows:

	Hectares
Om Saber	1,039
Salladdine	1,251
Ahmed Orabi	1,207
Western Tahrir	2,643

Arable and vegetable farming:

Eastern Tahrir	2,782
Northern Tahrir	3,029
Fatah Tahrir	2,284
Fatah I	2,327
Fatah II	2,172
Fatah III	1,951
Al-Intilaq I	740

In addition, there are 400 hectares in the Kitar area and 14 on the Cairo–Alexandria desert road.

The establishment of the Land and Water Research Centre, which was furnished with the latest technological equipment and manned by first-class experts, has contributed a great deal to the expansion and cultivation of the area, raising the fertility of the land and greatly enhancing its productive capacity.

Since the cost of reclamation and cultivation is relatively high, it was necessary to grow high-yield crops, mainly fruit and vegetables, on the reclaimed land. It is for this reason that special emphasis was laid on the planting in the southern sector of orchards, beginning with an area of 240 hectares in 1955, and continuing until 12,000 hectares were under cultivation, thus becoming one of the biggest centres for the production of citrus fruit in the whole Middle East.

Irrigation Methods in the Southern Sector

Irrigation by inundation

This system is used over an area of 10,000 hectares in the southern sector.

Irrigation by sprinkling

This means is used over an area of 21,000 hectares. It saves about two-thirds of the quantity of water needed to irrigate the same area by inundation. Moreover, it cuts down the cost of reclamation and allows for the cultivation of the entire reclaimed area without wasting parts of it on irrigation canals and ditches.

Animal products play a large part in the increasing development of Tahrir District. Animal husbandry stations are modelled on the most up-to-date establishments of their kind in the world, with canning factories as useful product outlets for preserving vegetables, fruit and jam.

Socially, enormous changes have taken place in the community life of the people of the new province. The farmers who came to this area from overcrowded provinces were given title to the reclaimed land at the rate of 2 hectares per family, as

The sprinkler system of irrigation gives greater mobility and economy. Pipes can be placed where required.

Cattle feeding on berseem (clover) after reclamation of land.

well as a modern furnished house and one cow. The population of this new district is now over 65,000.

Tahrir District is only one of the many similar land-reclamation projects which are being implemented in different parts of the country, and is typical of its kind.

At this point, however, it is worth pointing out that the average annual rate of land reclamation before the 1952 Revolution never rose above 1,000 hectares. Reclamation was undertaken almost entirely by big landowners to serve their own interests. However, although more intensive land reclamation operations began very soon after the Revolution, when the Tahrir District was first established, the stage of really large-scale scientific planning of land reclamation projects began in 1960/61, which heralded the first five-year plan.

The second five-year plan, covering the period from 1965/66 to 1969/70, showed a pronounced trend in the reclamation sector, based on the water resources currently available and on the additional resources which could be obtained. The plan also took into consideration the resources that would become available after the completion of the Aswan High Dam. Indeed, one of the most important benefits of the High Dam was the provision of enough water for the reclamation of about 520,000 hectares of new land, as well as making available some 730,000 hectare-metres of water for irrigation and the conversion of 394,000 hectares from basin to perennial irrigation.

The overall effect of the efforts exerted in this direction from the 1952 Revolution until the year 1969/70 was the reclamation of an area of 362,000 hectares for irrigation. Most of this is situated in the valley of the Nile and irrigated either by Nile water or from artesian wells; likewise the desert in the Nile Valley and Wadi Natrun; while on the north-eastern and western coasts the land is irrigated by rainfall.

The Process of Land Reclamation

So far we have been looking mainly at some facts and figures of land reclamation. Let us now consider what processes are involved and how they are planned and carried into effect. Reclamation entails the complete development of the area, beginning with the selection of the land in the first place, and going on to the study of the long-term economics of its reclamation and exploitation, as well as the more immediate priorities of the main irrigation and drainage projects and the cost involved. Work then begins on levelling the land to pre-determined elevations, on constructing the internal network of irrigation and drainage canals and carrying out civil engineering work, such as the building of barrages and bridges. This work goes on at the same time as the construction of pumping-stations, roads, houses and implementation of public services, and the laying of drinking-water pipes and provision of electricity.

In fact, the whole process can be divided into four main stages:

Preliminary exploration and research.

Planning, designing and preparation of the individual projects that together make up the whole scheme.

Implementing the projects.

Settling the people into their new environment.
Let us look at each of these stages in a little more detail.

Exploration and Research

Reclamation projects and the need for new land are preceded by a whole range of studies, including the following:

A geodetic survey which itself is carried out in three stages:

Exploratory Geodetic Survey

This is done by means of aerial photography, thus making it possible to eliminate from the beginning uncultivable land such as rocky outcrops and sand dunes. In this way much of the initial work is simplified, since efforts can then be concentrated upon the more potentially productive areas.

Semi-detailed Geodetic Survey

After eliminating the uncultivable regions, teams of international technical consultants visit the more promising areas to conduct surface examinations of the soil, to map the region in detail from aerial photography and to divide it into plots of about 100 hectares each. They take samples of the soil for analysis in special laboratories to check on its chemical properties, its composition, structure and porosity, and the percentages of the various saline elements that it contains.

Detailed Geodetic Survey

In the light of the results of the first surveys, a final selection is made of the most suitable areas for reclamation. These areas then undergo a more rigorous examination, after they have been subdivided into smaller plots of about 6 hectares each. Samples of soil from each of these plots are then taken for thorough analysis of their distinctive properties, before a final map-survey of the area is prepared, indicating the fertility classification and specific qualities of each plot. These maps, of course, become the bases for planning and implementing the individual projects.

Such a geodetic survey was completed for the High Dam area. Of six million uncultivated hectares covered by the survey, the following is a brief analysis of the types of soil:

Hectares

35,700	Class I	soil with high potential for a wide range of crops.
78,300	Class II	soil with good potential (includes fine-textured soils, heavy clays, loamy sands, and coarse sandy loams).
230,000	Class III	not good soil. (Shallow to rock; covered often with wind-blown sand; high water-tables without being irrigated; coarse-textured, gravelly and loamy sands not suitable for the efficient use of irrigation water. All lie at elevations that range from 9 to 125 metres above the river.)

[Source: *High Dam Soil Survey* (UAR General Report FAO/JF)]

The total area of these three classes of soils is about 344,000 hectares. This is about 137,500 hectares short of the land needed for the 885,000 hectare-metres of water available until 1977.

It may thus be said that the area of good land in this sample for living and crop growing beyond the present cultivated area is relatively small. The land available is poor (as indicated in Class III) and requires much time and effort before it can be made economically productive. Thus the problem here is not water but the availability of good land.

In addition, an aerial geodetic exploratory survey has been made covering an area of 21·4 million hectares in the Western Desert. It has thus been possible to delineate certain parts of it that are reclaimable, and research is now in progress into methods of irrigating these parts economically.

Topographical Survey

After the completion of the geodetic survey and the selection of the areas for reclamation, the places concerned undergo a topographical survey involving mapping the physical characteristics of the area and carrying out a formal land-survey.

Surface examination of the soil.

Planning, Designing, and Preparation of Projects

This second stage entails drawing up a complete plan for the reclamation and redevelopment of the area. Again, several major processes are entailed, including:

Designing the complete network of irrigation and drainage canals, together with their vertical and horizontal sections and inclinations, costing, and assessing the material and labour requirements.

Dividing the land into plots according to their distinctive topographical characteristics so as to ensure the unimpeded flow of irrigation water over the entire area.

Defining the size of villages and designing the public utilities and roads.

Defining the technical specifications and conditions required for the implementation of the different projects.

Estimating the manpower required for their implementation and establishing training centres to familiarise workers with the installation, maintenance and use of different kinds of machines and equipment.

Implementation

When planning and designing are over, detailed programmes, complete with time-tables, are prepared to co-ordinate all aspects of the project as a whole. The actual execution begins with the construction of the subsidiary roads that will link the reclaimed area with the general road network. This is followed by the digging of irrigation and drainage canals and by carrying out all the related civil engineering works: bridges, barrages, weirs, culverts, pumping-stations, etc. All this goes hand in hand with levelling the land, dividing it into fields and preparing it for cultivation.

Rehabilitation, Housing and Services

Land reclamation schemes do not have as their sole purpose the expansion of cultivation: they are also intended to provide a means of creating integral societies which enjoy all the amenities necessary for a reasonable standard of living. Reclamation projects, therefore, include new towns, villages, public services, buildings, hospitals, schools and agricultural co-operatives.

CULTIVATION

Putting the reclaimed land under cultivation takes place in four stages:

Research

Treatment and improvement of the soil

The 'sub-marginal' cultivation stage

The 'economic' cultivation stage

The Research Stage

Before putting the newly-reclaimed land under cultivation, a whole new range of analysis is made into the properties of the soil to determine the most efficient and economical means of treating and improving it in the shortest possible time.

For this purpose, experimental fields are established in various parts of the reclaimed areas as examples of the different categories of soils within it and of the

Dakhla Depression threatened by engulfment of the ever-present sand sea. The descent to desert can be halted and the contribution of these great regions to national incomes be made a reality.

Weathering by wind and sand has hollowed out the rock strata, often weakened by the existence of underground water.

environmental conditions prevailing there. To develop crop varieties best suited to the available land takes time; to re-establish grass on barren, eroded land again takes time. Experimental cultivation of improved types of crops, fodder, fruit-trees, etc., is then carried out in order to decide which variety is most suitable for cultivation in each particular spot.

Research is also undertaken to determine the average fertility-rating, the water requirements, the most appropriate irrigation intervals and the best means of nourishing the land in order to reach maximum production, as the content of organic matter in the soil is not high. The results of all these studies are incorporated into the technical recommendations which are finally formulated for the management of this type of scheme.

The Treatment and Improvement Stage

Here the properties of the soil are improved by mechanical, chemical, and biological means. Naturally, methods of treatment and improvement vary according to the nature and type of soil: for instance, the soils of the Nile Valley and Delta have a high clay content, and are rich in natural mineral nutrients vital for plant growth.

Saline soils, which contain a high percentage of damaging but soluble salts, are treated by 'washing' or flashing, that is by repeated inundation and dilution through the drainage network until the percentage of salts is reduced below the level that would be harmful to crops such as cotton. This process, of course, obviously depends upon an already adequate supply of water.

Sandy and gravelly soils are permeable, and do not easily retain water or fertilisers. They thus require special treatment by the addition of large quantities of organic matter and silt (calcareous alluvium).

The Sub-Marginal Cultivation Stage

This may well be the most important stage in the whole process of putting the reclaimed land under cultivation. It consists, quite simply, in cultivating the land with such crops, and by developing such varieties as seem best suited to its nature, while continuing to analyse the soil at intervals to ensure the progressive improvement of its properties by the introduction of nitrogen, phosphorus, etc.

The Economic Cultivation Stage

This is the last stage in the process of cultivating the newly-reclaimed land, and the one at which the land realises an economic return. This depends entirely on the selection of the most suitable crop or crops for each particular type of soil, and on enriching the soil and controlling irrigation, drainage, and so on. The major considerations which are taken into account in planning the cultivation schemes may be summarised as follows:

Extending the cultivation of green fodder crops with the main object of improving and building up the soil by application of plant nutrients, especially in sandy areas.

Prevention of erosion by wind and by excessive rain run-off. Stubble kept

64

above the soil surface cuts down surface wind speed and reduces soil blowing. Windbreaks of trees help, though these need water.

Selecting improved varieties of crops that can withstand high rates of salinity, such as berseem (clover) and alfalfa (lucerne).

Diversifying the crops in relation to their water requirements to ensure that the water resources available in a given area will suffice to cover the annual rotation cycle of that area.

Extending the cultivation of food crops such as cereals, oleaginous seeds and sugar crops wherever possible.

Cutting down the loss of grain production by the control of weeds is as important as the control of erosion.

Extending the cultivation of fruit, especially grapes and citrus, in limestone and sandy soils.

All this indicates, I hope, something of the strenuous efforts and the expenditure of time and money that are needed to bring newly-reclaimed land to the point of giving an economic return.

An ingenious method of protecting crops from damage by *Khamsin*.

These wind-breaks help prevent sand encroachment, as do the roots of trees and shrubs.

Herds of goats foraging for food on the fringes of cotton fields.

THE ORGANISATION OF LAND RECLAMATION SCHEMES

Land reclamation and cultivation schemes are carried out in Egypt by a group of organisations, corporations and specialist companies under the aegis of the Ministry of Land Reclamation. These are:

The General Land Development Organisation.

The General Desert Development Organisation.

The General Egyptian Land Reclamation Organisation, and its affiliated companies.

The General Egyptian Exploitation and Development Organisation of Reclaimed Lands.

We will now look at the work and responsibilities of each of these organisations in turn.

The General Land Development Organisation

This Organisation was founded in 1954; but it was soon expanded, when the first Five-Year Plan (1961–65) was drawn up, in order to oversee the extensive reclamation work envisaged as a result of the construction of the High Dam. The Organisation's task was to supervise and control the land reclamation aspect of the expansion, including the research work necessary before the policy for reclamation and development could be drawn up; and to plan the detailed schemes and supervise their accomplishment. This monumental assignment entails the following specific responsibilities:

Drawing-up the blueprints and the detailed designs of all reclamation projects, including land-levelling, irrigation and drainage canals, pumping-stations, etc.

Co-operating with the various Ministries and companies concerned in planning and designing the public utilities, such as electricity networks, aqueducts and the laying down of drinking-water pipes to the sites of projected villages.

Planning and designing the villages of the reclaimed regions and the homes of the farmers who will live there.

Linking the reclaimed areas to the general network of roads, and constructing the internal network of subsidiary roads within these areas.

Designing and erecting agricultural buildings, livestock stations and granaries, silos, refrigeration plant, etc. for the crops.

Designing and supervising the construction of public buildings—hospitals, schools, co-operatives, mosques, post-offices and police stations.

Preparing the detailed specifications of the various schemes, inviting and accepting local or international tenders for their execution and supervising their implementation.

Formally 'receiving' the reclaimed lands after the soil has been purified, having made sure that all the work entailed has been properly carried out and that all installations are functioning efficiently.

Handing over the reclaimed areas to the Exploitation and Development

Organisation of Reclaimed Lands so that the area can be placed under cultivation and continue to be supervised, maintained, and so on.

Up to the end of the financial year 1969/70, this Organisation had undertaken the task of reclaiming and rehabilitating a total of 273,000 hectares. The areas reclaimed are located in different provinces throughout the country.

The area of land reclaimed between 1952 and the middle of 1973, which excluded 880,000 feddans of basin land, was approximately 979,000 feddans. The aim was to raise the total amount to some 1,299,000 feddans, with the emphasis given to priorities such as the partial draining of Lake Manzala to improve 600,000 feddans of land. The Delta land and Upper Egypt, because of their obvious production potential, have received the largest share of aid for drainage schemes.

The General Land Development Organisation has also established and fully equipped one of the largest training centres of its kind in Egypt and staffed it with personnel expert in all relevant fields, in order to create a whole new generation of trained and experienced workers to operate and maintain the various items of machinery and equipment used in the reclamation of land and putting it under cultivation.

The General Desert Development Organisation

This Organisation was founded in November 1958 to contribute to the development of the national economy, mainly in the sphere of the reclamation of the desert outside the Nile Valley. To this end, the Organisation has been concerned with conducting research on ground-water, with carrying out a complete survey and classification of desert lands for water-development potential, with studying plant and animal production, and with the other schemes necessary for the creation of integral and prosperous societies in desert areas.

The land in general beyond the Nile Valley, that is the vast expanses of desert, contain relatively few areas which are potentially able to be developed because of either soil conditions or lack of water; although there have always been from ancient times oases where springs or wells have provided water for animals and men.

The target was a total of 1,299,000 feddans by 1975, with priority being given to such developments where there is greater potential, such as the New Valley Project.

The 'New Valley' Project

This is one of the largest schemes in the world in which land reclamation depends on ground-water, considerable amounts of which make it independent of

The fruits of land reclamation and irrigation.

irrigation from the Nile. The region is underlain by Nubian sandstone at different levels and comprises the oases of Kharga, Dakhla, Farafra, Behariya and Siwa, which form depressions extending across the Western Desert from south-east to north-west. (The search for ground-water has so far been confined to the Western Desert.)

	Area irrigated
Dakhla Depression	30,000 feddans
Kharga Depression	20,000 feddans
Farafra Depression	100 feddans

69

Irrigation by gravity flow.

A traditional way of lifting water for irrigation by *sakias*, in the Faiyum Depression.

Since 1959, the General Desert Development Organisation has undertaken a series of studies, in co-operation with international companies and with prominent experts from the United Nations, in order to arrive at the best economic and technical bases for extending this project by using the underground reservoir in the Western Desert. The most up-to-date apparatus, including a specially-designed computer, was used to find out how much water was stored underground and the area which could be economically cultivated using this supply.

So far, an area of about 200,000 hectares is now being cultivated, depending for irrigation on artesian wells. It is suggested by some experts that there is a possible potential of 40 million hectares, 20 per cent of this water being fairly easy to obtain in the New Valley, but high cost may preclude the attainment of such a large area of reclamation. The reclaimed area, for example the Kom Ombo Valley in Upper Egypt, was provided with all the housing and social amenities recommended by the units researching into the resettlement of desert communities, such as the 50,000 Nubians who were forced to leave their land because of the formation of Lake Nasser.

The Wadi Natrun Project

The area concerned lies to the north-east of the Western Desert, half-way along the Cairo–Alexandria desert road. The wadi, formerly a saline depression, is 61 kilometres long and 10 kilometres wide. This smaller project has brought into being 10,000 hectares which depend for irrigation on water pumped from artesian wells.

The North-Western Coast

This area, which extends along the coast for a distance of 523 kilometres from Alexandria to Salum, is inhabited by a mainly Bedouin population of about 100,000. Their main occupation is raising sheep on the scanty grazing potential of the land; but most of them also tend small vineyards and orchards of figs, olives, and almonds, which depend on the winter rains that amount on average to 150 mm per annum. Forage is poor in the mountains east of the Nile Valley.

The General Desert Development Organisation has also carried out extensive studies of the natural potential of this area with the aid of the United Nations Special Fund and the Food and Agriculture Organisation, and hopes to draw up a comprehensive development plan for the whole region. Meanwhile, the Organisation has carried out in the past few years some experimental projects in an area of 6,750 hectares, using the water which floods the valleys during the rainy season and other ground-water to irrigate experimental orchards. This constitutes a part of the Government's plan to give the Bedouin population a more settled way of life.

The North-Eastern Coast Projects

The Organisation has been successful in using the rain-water in the Sinai coastal region to reclaim and put under cultivation an area of 3,750 hectares, complete with the necessary villages, public services, roads and electric power supply.

The Bitter Lakes Project

The Organisation's activities extended to the heart of the Sinai Desert where a complete survey was made to determine and classify reclaimable lands. In the light of these studies, the Organisation began in 1964/65 to reclaim an area of 8,000 hectares in the region of the Bitter Lakes, east of the Suez Canal. For the first time in history, Nile water was to be carried to the heart of the Sinai Desert through a culvert which was constructed under the bed of the Suez Canal.

Work had begun on the irrigation of an area of 800 hectares, and the civil engineering work had almost been completed, when this unique enterprise was interrupted by the war of June 1967.

The Maryat Project

The area covered by this project is about 27,500 hectares. All reclamation and development schemes in connection with it are now complete, including all the amenities necessary to a modern society. In fact, 20,000 hectares of this area have already been handed over to the General Egyptian Organisation for the Exploitation and Development of Reclaimed Lands.

The Minya Project

The Organisation has also reclaimed and turned over to the Exploitation and Development Organisation an area of 1,600 hectares to the east of the Nile, in the province of Minya.

The General Egyptian Land Reclamation Organisation

This Organisation was established in December 1961 to take over and coordinate the activities of various companies which were engaged in land reclamation.

In March 1964 the Arab Company for Land Reclamation was created by Republican decree and affiliated to the Organisation to specialise in reclaiming scattered areas of uncultivated land in the Nile Valley.

The Ministry of Irrigation, in indicating the priorities in development expenditure for 1966/67, gave high priority to the conversion from basin to perennial irrigation, employing just under half the Government's funds; 17 per cent was allocated to reclamation, 13 per cent to general irrigation, and 8 per cent to covered tile drainage. Only about 5 per cent was given for new irrigation projects, improvement and maintenance of pumps (replacement included), and additional water resources. Prior to and including 1966, practically no finance went into research. The general trend has been for an increase in reclamation because of the additional water made available by the completion of the High Dam. Nevertheless the change from basin to perennial irrigation does not on the whole change the annual demand for water.

Two special government agencies, the High Dam Authority and the Aswan Regional Planning Project, supervise all aspects relating to the Aswan High Dam.

In the important field of scraping and levelling, the Organisation's companies

The cultivation of special deep-rooted grasses is one of the early stages in land reclamation. In desert areas experiments are being made with acacia seeds planted in goat droppings. When the rains come, the seeds germinate and eventually the trees will not only act as effective wind-breaks but will help to stop sand encroachment.

The cultivation of seedlings in a protected nursery environment in Tahrir District.

were concerned with the improvement of irrigated land by the process of levelling it with their fully mechanised 'fleet of equipment', which in turn meant that water was applied to the land uniformly—an important tenet in irrigation for high yield of crops.

In contrast to levelling is a need for an exact grading of land in larger areas, so that furrow irrigation rather than flood irrigation can be used, with a consequent saving of water.

The building of the barrage meant that canals could have water in them all the year round, except for a mid-winter period set aside for cleaning and repair and maintenance in general. With the abundance of water available by gravity and not charged according to the amount used, overwatering will almost certainly occur. The Delta is no exception to this rule: high water-table and saltiness have resulted in seriously reducing crop yields.

So as to conserve water and reduce the hazard of a high water-table, the Egyptian Government has tried to stop the practice of over-use of water by the farmer (local 'ethics', according to Clawson, 'rule out water charges'). Control is carried out by regulating the size of the field outlet and the size of the head on it; this in turn regulates the total flow into the branches and delivery ditches. In the past, irrigation canals were designed to be 250 mm above the ground elevation, but this led to over-use of water. In the new schemes, and when basin to perennial conversion was made, the operating level is some 500 mm below the surface where the farmer has to receive it; often a lifting device such as a screw, pump, or water-wheel has to be used. The amount of labour required in lifting thus acts as a system of water rationing.

The General Egyptian Exploitation and Development Organisation

This Organisation was established in 1965 to take charge of the further development and exploitation of the increasing areas of land that were initially reclaimed and put under cultivation by other organisations.

It was evident from the beginning that the Organisation's mounting responsibilities would eventually extend to the cultivation of an area of some 5–600,000 hectares by 1980 and to the development on socialist principles of the new societies to be composed of the inhabitants of the reclaimed regions, grouped into about 500 new villages. The fulfilment of all plans for all cultivated land would involve over 3 million hectares.

The lands turned over to the Organisation for cultivation and development were divided on geographical and topographical lines into nine sectors, ranging in size from 21,000 to 50,000 hectares. These sectors were in turn divided into farms averaging about 2,000 hectares each. The farm was the basic unit in the whole system, larger farms possibly being formed from aggregates of smaller units prior to land reform. However, the size of small farms has been increasing since 1950 (land reform has tended to accelerate this). About 75 per cent of the labour force, consequently, is concentrated on the smaller farms, 85 per cent of all holdings accounting for less than 40 per cent of the area covered.

74

Soil and Water Research

The Organisation has established a highly-specialised technical capacity for taking over the newly-reclaimed lands, for conducting the soil and water research work in each area, and for drawing up a complete programme for future development. One of its main concerns is to devise means of satisfying the water requirements as the properties of the soil improve. Irrigation by sprinkling was found to save more than half the water previously needed to irrigate sandy soil; and the Organisation can now boast of having under its control the biggest area of land irrigated by sprinkling in the Middle East: about 34,000 hectares in all.

Agricultural Production

The total area of land reclaimed by the Organisation and incorporated into the regular agricultural cycles in the year 1969/70 was 170,000 hectares. The main summer crops produced in these areas in 1969 were rice, cotton, maize, groundnuts and sugar cane, in addition to berseem (Egyptian clover) as animal-fodder for raising livestock and for producing the necessary organic fertilisers for the improvement of the soil by the use of nitrogenous fertilisers. Prior to the construction of the High Dam, the Nile brought organic matter which added to the fertility of the soil. All desert regions are low in nitrogen content, and not sufficient humus has been added to the soil over the centuries. This type of fertiliser is used basically on all irrigated land, although it is not required for rain-fed areas on the whole. Crop production dominates the agriculture, with livestock production taking second place and forming about one-third of the Egyptian GNP.

The Organisation is also greatly interested in expanding the area of orchards and vineyards, which now total 10,000 hectares. The vine, like the olive tree, is deep-rooted and can thus draw water from deeper levels. There are 3,600 hectares of vines, 4,800 of citrus and mango trees, 1,300 of olive and other fruit trees, and 500 of orchard nurseries. Indeed, orchard and vine production in the newly-reclaimed lands is proving increasingly successful. The Organisation has accordingly drawn up a programme aiming at increasing the area of vineyards, where there is chalky soil in the reclaimed areas, at the rate of 2,000 hectares per annum up to a total of 15,000 hectares; and also at increasing the area of citrus fruit in areas of sandy soil at the same rate, to reach an ultimate total of some 20,000 hectares. Stress is being laid on the cultivation of certain varieties of citrus fruit such as oranges, for which there is special demand in foreign markets.

Livestock

The Organisation is very keen on developing animal husbandry in the newly-reclaimed areas to serve the following purposes:

To consume the green fodder crops that are being widely cultivated because of their important contribution of nitrogen to soil-improvement. Berseem clover occupies the largest area of any crop in Egypt.

To acquire animal manure as the organic fertiliser which is still much needed to improve the properties of the soil, in spite of the wide use of chemical fertilisers.

To increase the quantity and improve the quality of meat and dairy products. Berseem is a highly nutritious feed.

The Organisation now keeps 40 per cent of the four million head of breeding cattle to be found in the Middle East. These are kept on farms, as the desert cannot provide feed for cattle. The majority are pure Friesian or cross-bred, and the re-

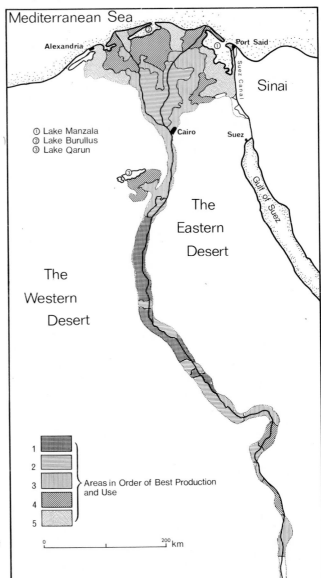

The use to which fertile or poor land is put is not governed by the density of population, as might be supposed. In Upper Egypt, for instance, according to W. B. Fisher, 'the better land tends to be occupied by the heavier rural population'. In the Delta this is not so, even though the last two classes of area indicated in the key to the diagram occupy an even higher proportion of the agricultural population.

mainder of local breed. The Organisation is also interested in sheep, 67 per cent of its flocks being either pure or cross-bred Merinos. The herding of domestic livestock depends on available feed supply, which in turn depends on the increase and decrease of precipitation. In Egypt there has been no consistent trend in sheep numbers over the last thirty years.

The Mechanisation of Agriculture

The Organisation's policy tends towards expanding the use of modern machines in all processes of land treatment, pest control, and agriculture wherever this is possible; the control of weeds on rain-fed and irrigated crop areas is of major importance if what has been gained in soil fertility and added moisture is not to be lost. Especially in the cultivation of crops such as wheat, barley and maize is the use of machines encouraged, thus achieving up to 50 per cent saving in seed and 20 per cent increase in crops. Moreover, mechanised agricultural methods have reduced production costs by an average of £E 16 a hectare per annum.

Rehabilitation and Development of Society

Over 220 new villages have been established in the 30,100 hectares of newly-reclaimed areas which the Organisation controls. They are inhabited by some 28,000 families, comprising about 140,000 men, women and children. The land was given to these families either freehold or under lease, under the Government's land reform policy.

As well as the houses and all the public services provided for these villagers, the Organisation has given the engineers and workers employed on the various projects and their families adequate housing and all the necessary health, educational, cultural and entertainment facilities.

Other institutions established in these areas to help in the development of the new communities on a co-operative basis include the following:

Rural development societies, formed by the beneficiaries themselves, with the object of promoting rural and environmental industries, home economy, afforestation, family planning, and so on.

Agricultural co-operative societies which have a total membership of more than 12,500. Although agricultural co-operatives had been in existence before land reform, many more came into being as a result of it, and the aim is to promote these throughout the country. The agricultural co-operatives serve as practically the only means of obtaining credit and supplies, and as the only outlet for the marketing of cotton. In essence the farmer manages with the supervision of an employee (governmental). In this way the Organisation or the Government deals with the co-operative through the medium of the farmer.

Consumer co-operative societies, at the rate of one per sector, under the Organisation's control, with different branches located in such a way as to ensure a ready supply of consumer goods to all members of the communities, have also been established.

The Institute of National Planning

This Institute was established towards the end of 1968 in order to enlist the services of scientists and other experts to give technological advice and to help solve the many problems which arise in the process of reclaiming and cultivating new and potentially new land. Members are top-ranking University professors and other prominent experts in this field, together with representatives of the various ministries, organisations and companies concerned. Its terms of reference include: the preparation of expert studies on all programmes of reclamation and on the use of new land; monitoring the progress of the soil programmes and recommending ways of overcoming the difficulties that may be met in the process; and co-ordinating the work of the different ministries, departments and organisations concerned. To this end, the Institute has formed a number of specialised commissions, covering the following activities:

> Land and water allocation: the availability of water for irrigation is controlled by the Ministry of Irrigation
> Field crops
> Orchard crops
> Livestock, including improvement of breeding by good sires, re-seeding, grazing rotation
> Economic studies
> Engineering, irrigation and drainage construction
> Agricultural machinery
> Underground water

Members of these commissions visit the reclaimed areas, conduct their researches and formulate their recommendations for solving whatever problems they encounter.

Irrigation Developments

Two of the major land reclamation schemes are west of the Nubariya Canal and in the Huseiniya Plain.

The first is an area extending as far as Dabaa. According to the survey of the land resources of Egypt, this area ranks among those best suited for reclamation in the whole country. This was one of the many considerations which led to this scheme being given priority. It aims at achieving a significant expansion of the arable area by using the extra Nile water now stored behind the High Dam. Because of this storage, Egypt will be able to irrigate 530,000 hectares of land which until then had been unreclaimed. Rice output will thus be able to be increased. Perennially irrigated via a major new canal, named the Nasser Canal, with outlet off the main course of the Nile downstream from the Delta Barrage, the crop-scheme designed for this project includes the introduction of sugar-beet as a major crop for the first time in Egypt's history. It also looks to a considerable expansion in the production of vegetables, fruit and oleaginous crops, as well as cotton, maize, wheat and clover.

The reclamation in the Huseiniya Plain and the area south of Port Said extends from Ismailia to Port Said. The scheme was planned to include the construction of urban centres, villages, power grids, pumping-stations and roads; but its completion was interrupted by the June War of 1967.

THE LAND AND THE PEOPLE

The population of Egypt is estimated, as we have seen, at about 34 million, of whom a mere 350,000 live outside the Nile Valley and the Delta. This small fraction of the population constitutes the total population of the desert areas, namely, the vast expanses of the Western Desert, the Eastern Highlands and the Sinai peninsula—which occupy 96·2 per cent of the total area of the country. In other words, 98·8 per cent of the Egyptian population live in an area not exceeding 3·8 per cent of the entire area of the country. Thus, *habitable* Egypt is in fact one of the most densely populated territories in the world, since the average per capita share of land is o·11 of a hectare if we take into account the entire area of the Nile Valley and Delta, and o·08 hectares if we include only the actual cultivated area. This compares with an average o·36 hectare per capita share in Europe and a 1·58 hectare share in the United States (see graph on page 35).

Indeed, the most crucial problem of the Egyptian countryside is the comparatively small size of the cultivated area; a problem which is further aggravated by the fact that the rural population is increasing at an annual rate of nearly 3 per cent, whereas the cultivated area is increasing at no more than o·5 per cent.

It is estimated that between 1952 and the middle of 1973 some 979,000 feddans of land have been reclaimed. In addition 880,000 feddans of basin land have become viable. With the increase in population and the pressing need for higher output, the Delta land, being more fertile, has top priority. Lake Manzala is in the process of being drained, which will contribute to the improvement of about 600,000 feddans of land.

This situation is the reverse of that which prevailed early in the last century, when Egyptian agriculture was hampered mainly by a shortage of manpower. And in spite of a tremendous increase in the population which began towards the middle of the century, Egyptian agriculture continued to suffer from this shortage for a long time. In the opinion of some scholars, this state of affairs was due to a constant expansion of the arable area and of summer cultivation which required an ever-increasing number of hands to raise the irrigation water from the river level to their fields, often by crude methods using buckets and primitive levers.

Until the early part of this century, the cultivated area continued to increase at a rate compatible with the increase in the population. But the expansion of the cultivated area began to level off, while the population continued to increase at an even higher rate than ever before, owing to the improvement in the health services which cut down the death-rate and raised the birth-rate. And although improved irrigation and drainage services and the introduction of modern methods of agriculture have helped to increase the area under crops, the average per capita share of the cultivated area has continued to fall because of the disproportionate increase in the population. Thus, the expansion of the cultivated area has now become the most crucial problem which has to be faced in order to meet the ever-rising demand for foodstuffs from a constantly increasing population. Official Egyptian sources have now divided the areas of cultivation into five classes or zones, each according to productivity capability.

Another major problem which impeded agricultural progress in Egypt was the

Agrarian reform provides standardized houses for farm workers, within easy reach of their fields.

A Bedouin ploughing land.

unequal distribution of the cultivated lands before 1952. The statistics shown in the table on p. 82 would suggest that the reform laws have radically changed the distribution of land ownership in Egypt: small ownerships have increased by 15 per cent (or by 75 per cent if area is taken into account); the largest landowners now have much smaller estates. The middle-sized group has increased by about 15 per cent (25 per cent in land area). Thus there has been a shift in power from a very large section of big landowners to a larger group of small owners, with often the choice of land distribution being awarded on welfare grounds rather than efficiency, with minimum units, often less than one hectare, being allotted to many.

One of the six principles proclaimed by the 1952 Revolution was to 'eliminate feudalist tyranny over the land and over those who live on it.' Hence, the first concrete step taken by the Agrarian Reform Authority in the sphere of economic and social reform was the promulgation of the Agrarian Reform Law No. 178 of 1956, later amended by Law No. 127 of 1961, which had the following major objectives:

> To reduce the power of the large landowners by reducing the maximum holdings that any family could possess to 200 feddans, which in turn were further reduced to 100 feddans by the second land Reform Law of 1961. Land ownership was still further curtailed in 1969, when a limit of 50 feddans per family was imposed. Land in excess of this limit, amounting to about 400,000 hectares, was requisitioned and redistributed to former tenants and landless farmers.

> To regulate the relationship between landowners and tenants, for which purpose the Law stipulated that land might not be let except to the farmers who actually worked it, and fixed the rent at seven times the tax payable by law. In this way the exploitation of small tenants by middlemen, which used to be the common practice, was abolished. In the case of sharecropping tenants, the law fixed the landowner's maximum share at 50 per cent after deducting all expenses. Redistribution of land by 1966 affected 960,000 feddans, and 4 million feddans were affected by the new tenancy laws. This amounted to 17 per cent of the former and 65 per cent of the latter.

> To safeguard the rights of agricultural workers by appointing a special commission to fix their minimum wages from year to year and giving them the right to establish their own trade unions for the first time in Egypt's history.

> To organise agricultural co-operation through the establishment of cooperative societies among small landowners to give them credit and to supply them with seed, fertilisers, livestock, machinery, storage and transport facilities.

Meanwhile, the Agrarian Reform Authority embarked upon costly programmes for agricultural expansion. Investments allocated to the agricultural sector in the first Five-Year Plan for Economic and Social Development (1960–65) amounted to about £E 347,000,000, representing 20·5 per cent of the total investment allocated to the entire plan. The implementation of this plan brought about an increase of £E 123·3 million in the cash income from agriculture, that is, from £E 405 million in 1959/60 to £E 528·3 million in 1964/65.

Land Holdings and Total Land Area by Three Size Groups

	1952 [a]	1952 [b]	1961 [c]	1965 [d]
Number of Holdings	('000 hectares)			
less than 2 hectares	2,642	2,841	2,919	3,033
2–42 hectares	154	162	182	178
over 42 hectares	5	5	0	0
Total	2,801	3,008	3,101	3,211
Total Area	('000 hectares)			
less than 2 hectares	891	1,168	1,332	1,551
2–42 hectares	945	1,013	1,223	1,163
over 42 hectares	678	333	0	0
Total	2,514	2,514	2,555	2,714

a = Prior to Agrarian Reform Law.
b = After Agrarian Reform Law which set maximum ownership of 200 feddans (84 hectares).
c = After new Agrarian Reform Law which reduced maximum ownership to 100 feddans (42 hectares).
d = Excludes State-owned lands and Agrarian Reform lands not yet distributed.

Source: UAR, Central Organisation for General Mobilisation and Statistics, *Annual Statistical Abstract 1952–1966* (Cairo, June 1967).

Principal Crops

	Area ('000 feddans)		Production ('000 metric tons)	
	1971–72	1972–73	1971–72	1972–73
Wheat	1,246	1,249	1,618	1,838
Maize	1,538	1,656	2,421	2,508
Millet	483	487	831	853
Barley	97	86	109	97
Rice	1,146	997	2,507	2,274
Berseem	2,819	2,874	n.a.	n.a.
Beans	337	272	362	273
Lentils	67	74	54	62
Onions	49	45	490	530
Sugar cane	202	198	7,701	7,349

Source: *The Middle East and North Africa 1975–76*, Europa Publications (22nd edition).

But, although the average yield per acre in Egypt is higher than elsewhere in the Arab countries, it is still considered to be insufficient (three-quarters of the food-stuffs consumed), taking into account the Egyptian farmer's hard work and the exceptional fertility of the Nile Valley's soil. Owing to intense population pressure, it was thus necessary to devote special efforts to increasing arable land productivity in accordance with a carefully conceived plan for vertical expansion. This plan was based upon several objectives: the provision for the farmers of sufficient quantities of seed, fertilisers, insecticides and other production requirements at minimum cost; the mechanisation of agriculture to economise on human labour and animal labour and to ensure greater efficiency; the control of irrigation and drainage to achieve a higher intensity of soil cultivation; and the regulation of the agricultural cycle to reach a state of optimum production from the optimum unit.

Egypt has made satisfactory progress towards achieving these goals through promoting agricultural co-operatives, while the problem arising from the frag-mentation of landholdings was overcome through the introduction of community cultivation, thus paving the way for the mechanisation of agricultural processes and for reducing production costs per unit. The technical advantages of this form of cultivation is that large-scale farming methods can be applied to small plots (ploughing, irrigation, and crop spraying can be carried out simultaneously).

Another feature which jeopardised Egyptian agriculture for many decades was its dependence on cotton as the only major export crop. Significantly, Egypt is the major grower of the long-staple variety (40–50 per cent of the world's cotton output). Cotton is, of course, a very unpredictable crop since its price on the world markets is susceptible to wide and frequent fluctuation. Successive governments in Egypt invariably tried to induce Egyptian farmers to depend less on cotton and to encourage the allocation of more acreage to wheat and other cereals in order to meet the increasing demand on food; but these efforts were impeded by the lack of the irrigation water which was so absolutely necessary for diversifying agri-cultural production. However, the construction of the High Dam has at last put an end to this problem, and crop-diversification is now taking place rapidly, and the extent of its full use can now be evaluated. It has added one-third to the total of cultivated land in Egypt.

When cotton usurped wheat and berseem, there was not enough nitrogen for both cotton and wheat. Hence Egypt's widespread use of fertilisers.

With the completion of the High Dam, a rethinking of the best rotation of crops has become necessary. For instance, rice and maize have been planted where there was enough water for them—there is the possibility of a better maize yield if planted earlier in the year. The Dam has changed the pattern of agriculture quite extensively.

Nevertheless, owing to perennial irrigation there has been an increase in the saline content of the soil. This is because of more extensive evaporation from the soil, and has become an acute problem in areas where cotton growing is specialised. According to Issawi, drainage is of even greater importance than irrigation.

The diversification plan aims at encouraging the cultivation of rice, sugar-cane, sesame, groundnuts, flax, fruit and vegetables, and other, mainly food, crops which find a ready market abroad. The cultivation of rice, in particular, has undergone a vast expansion under the new plan. More than a million acres are now devoted to its cultivation every year, which means that three-fourths of all the rice in the Middle East is grown in Egypt. New varieties have been introduced, and it has become a major export. If rising production trends continue, not only in rice but in wheat and other cereals, then Egypt will be self-sufficient in the near future. This also applies to the high yielding crop of sugar-cane. The expanding sugar industry has greatly increased the demand for this. Moreover, the plan places great stress on animal husbandry as an important aspect of agricultural activity. In this connection, the Government is importing first-class breeds of cattle, sheep and poultry to improve the quality of local breeds, and every effort is being made to impress upon the Egyptian farmer that animal husbandry is an important and profitable aspect of his occupation.

A major step forward. The reopening of the Suez Canal in June 1975. President Sadat sailed in the ship which headed the convoy to mark the official opening of the waterway.

REFERENCES

Marion Clawson, Hans H. Landsberg, Lyle T. Alexander, *The Agricultural Potential of the Middle East. Economic and Political Problems*, Elsevier, New York, London, Amsterdam 1971

Peter Mansfield, *The Middle East. A Political and Economic Survey*, 4th edition, Oxford University Press, 1972

J. I. Clarke, W. B. Fisher, *Populations of the Middle East and North Africa. A Geographical Approach*, University of London Press, 1972

W. B. Fisher, *The Middle East*, 6th edition, Methuen & Company, 1971

Michael Adams, *The Middle East. A Handbook*, Anthony Blond, 1971

The Middle East and North Africa 1975-76. A Survey and Reference Book, 22nd edition, Europa Publications

H. E. Hurst and P. Philips, *The Nile Basin*, 7 volumes, London 1932-38

H. E. Hurst, *The Nile*, London 1952 and *The Major Nile Projects*, Cairo 1966

Gabriel S. Saab, *The Egyptian Agrarian Reform 1952-62*, Oxford University Press, London and New York 1967

Doreen Warriner, *Land Reform and Development in the Middle East. A Study of Egypt, Syria and Iraq*, 2nd edition, Oxford University Press, London 1962

The Democratic Republic of the Sudan

The Democratic Republic of Sudan, with an area of just under 2·5 million square kilometres, is the largest country in the continent of Africa and constitutes 22 per cent of the entire land area of the Arab World. It is bounded on the north by Egypt, on the west by Libya, Chad and the Central African Republic, on the south by Zaire, Uganda and Kenya, and on the east by Ethiopia and the Red Sea. From Wadi Halfa on the Egyptian border to Yambio in the extreme south it extends for over 2,250 kilometres, while from east to west at its widest point the Sudan covers almost 1,600 kilometres.

With an estimated population of only about 16,000,000, the Sudan is two and a half times as large in area as Egypt. It shares with Egypt the same two outstanding characteristics—its dependence on the Nile (or, rather, on the two Niles, the White and the Blue), and its vast stretches of desert. The topography of the Sudan is, however, distinctively different from that of Egypt, and its climate, water-sources and patterns of agriculture are more complex. It is therefore of importance to discuss the physical characteristics of the Sudan in greater detail in order to achieve some understanding of developments in irrigation and land use.

The south of Sudan enjoys, or endures, a tropical equatorial climate, with a long rainy season and an annual fall of well over 1,400 mm. Farther north a quite different climatic regime prevails; at Khartoum, for instance, the annual average rainfall is only 135 mm, while in the Nubian desert there is scarcely ever any rain at all. Temperatures can be very high in central and northern Sudan: Khartoum suffers about 33°C in early summer. At Yambio in the south, the average daily maximum temperature in February and March is only about 27°C, but the humidity is, of course, very much greater.

With the increased rainfall southwards, crops such as sorghum and bulrush millet are grown, and with the abundance of water available, groundnuts and sesame too. The speed of the cultivator in the various stages of preparation of the soil is of paramount importance in raising the crops.

From a point north of Port Sudan, linking points north of both Atbara and Khartoum and Jebel Marra (latitude 15°N) on the west, a boundary (isopleth) division can be drawn climatically which gives a northern zone of winter maximum rainfall, contrasting with most of the rest of the Sudan which has a summer maximum.

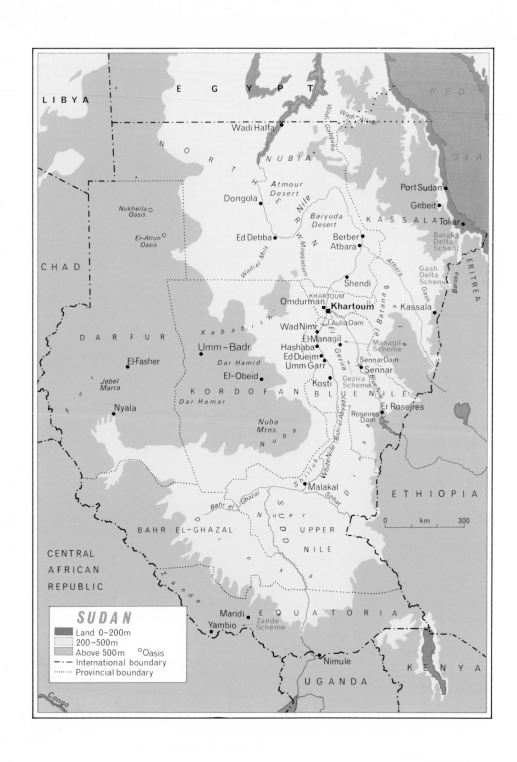

LIBYA

E G Y P T

RED

SEA

Wadi Halfa

N
O
R
T
H
E
R
N

NUBIA

Atmour
Desert

Dongola

Baiyuda
Desert

Port Sudan

Gebeit

KASSALA

Tokar

Nile

Ed Debba

Berber
Atbara

Baraka
Delta
School

Nukheila
Oasis

El-Atrun
Oasis

Wadi el Milk

Wadi el Maqadam

Shendi

Gash
Delta
Scheme

Atbara

CHAD

Omdurman

KHARTOUM

Khartoum

Kassala

Baraka

Gash

Kababish

Wad Nimr

J. Aulia Dam

DARFUR

Umm–Badr

El Managil

Hashaba

Ed Dueim

Managil
Scheme

ERITREA

El-Fasher

Dar Hamid

Umm Garr

Sennar Dam

Sennar

Jebel
Marra

El-Obeid

KORDOFAN

Kosti

Gezira
Scheme

B L U E

N I L E

Bahr el Abyad

Dar Hamar

Nyala

Nuba
Mtns.

N
u
b
a

BLUE NILE

Er Roseires

Roseires
Dam

0   km   300

White Nile

Shilluk

Malakal

Sobat

ETHIOPIA

Bahr el Ghazal

D

N
u
e
r

Gbazal

BAHR EL-GHAZAL

Azande

S
U
D
D

D
i
n
k
a

UPPER

NILE

CENTRAL

AFRICAN

REPUBLIC

Maridi

Zande
Scheme

E Q U A T O R I A

Yambio

Nimule

KENYA

UGANDA

Congo

Jebel Marra, the highest mountains in the country.

NATURAL FEATURES

The Sudan is predominantly a flat country bordered by mountains to the east and south with hardly any outstanding topographical features apart from the Nile and its numerous tributaries and a few rugged hills scattered here and there in an uneven pattern.

The central Sudan forms a mighty depository for the sediments accumulated by the main Nile. In fact, most of the river sediment is redeposited in the Sudan, forming silt and clay plains, after only a short travel and natural levées. (Most of the sediment which reaches Egypt is brought down by tributaries from Ethiopia.)

In the north, the desert stretches out on both sides of the Nile and there is little else to be seen as far south as Khartoum. This desert is in fact a continuation of the Great African Sahara. The area of Nubia between the Dongola Bend of the Nile and the railway of Wadi Halfa is known as the Atmour Desert. This is a sandy, gritty plain, and is one of the almost totally arid regions of northern Sudan. This inhospitable, arid area has formed for centuries a southern barrier to the territory of Egypt. The plain is broken here and there by low hills 30 to 300 metres in height; occasionally there are enough of these to be called a range. The only populated and economically viable regions in this area consist of a series of basins in the form of isolated oblong oases on the edge of the desert: these often reach down to the banks of the river. There is ribbon cultivation in the trough of the Nile.

89

To the east of this extremely arid region the topography of the land changes, gradually giving place to volcanic rocks which rise above the surface and form very steep ridges often exceeding 900 metres in height in the Red Sea hills. These features become more accentuated as we move eastwards towards the Red Sea, where they join the south-eastern highlands of Egypt and the Eritrean plateau which forms the western ridge of the Great Rift Valley in this part of Africa. Several creeks run down from this high plateau: some flow westward to the wadis of Allaqi and Gabgaba, while others turn east into the Red Sea. In the Western Desert region, on the opposite side of the Nile, the rainfall is very scarce indeed. But underground springs give life to scattered oases, such as, for example, Nukheila and El-Atrun. The south-eastern end of this arid region is known as the Baiyuda Desert. It is bordered on the east and north by the Nile and on the west by the Wadi el-Milk, which is a dry shallow valley ranging in width from 4 to 5 kilometres. It begins at a point close to Umm-Badr and ends at the bank of the Nile near Ed-Debba. Second in importance is the Wadi Muqaddam which begins some 100 kilometres to the south-west of Omdurman and runs for about 320 kilometres across the desert to end at the banks of the Nile at Kosti.

To the south of this desert region lies a wide plain of sandy soil, extending west-wards from the White Nile and passing through the Kordofan and Darfur provinces. The plain is intermittently broken by hills which rarely exceed 90 metres in height. It is also interrupted by a few shallow dry wadis which eventually peter out in the sand. The red-grey soil of this area is more fertile than that of the northern plains. Southwards, below latitude 14°, the sandy plains give way to the black soil which covers the central Sudan, extending from the southern end of Darfur province (where there is 'episodic drainage' and water percolation from the Jebel Marra) across the Nile Valley to the southern borders of Kassala. These central clay plains form a low-lying zone with shallow river valleys where flooding can be contained. Irrigation is possible. This area has the greatest population density in the Sudan. The eastern part of these plains, between the River Atbara and the Blue Nile, is known as El-Batana and is noted for the fertility of its soil.

In the angle between the Blue and the White Nile lies the Gezira Plain. This area is completely flat and almost devoid of any topographical features, so that it is very difficult to define the watersheds of the two rivers without making a full survey of the whole area. By contrast, however, the plains of Kordofan and Darfur are broken by several ranges of hills, and in the far west is the highest range of moun-tains in the whole Sudan, Jebel Marra, which covers an area 130 kilometres long and about half that distance wide. At its highest point it rises to about 3,000 metres above sea level, high enough to induce rainfall and giving rise to a number of seasonal streams that flow westwards. The semi-desert area has characteristically a zone of sand dunes in the middle, with a hummocky type of topography. This is known as the Qoz region. Within the clay basins there is cultivation, but in general the area is sparsely populated by nomadic tribes. Westward from this range runs the Wadi Ibra, the largest valley in western Sudan, which is filled with rain-water from the mountains for three months every autumn.

The Nuba Mountains occupy the south-eastern corner of the province of Kordo-

These curious holes, found in the Province of Kordofan, have been made by generations of Nuba women grinding grain.

fan, with ridges rising to a height of between 500 and 700 metres and some peaks reaching 1,000 metres. These mountains are surrounded by vast plains covered by sedimentary soil which varies in texture from coarse sand to fine clay. The sandy soil is called *el Gardud* by the local people, while the clay is simply known as *teen*, or mud. The Nuba people who live in these mountains have succeeded in building an intricate system of terraces on the mountain-sides. It is evident, however, that this system of cultivation was once more widely used than it is nowadays, since the Nuba people gradually abandoned the mountains for the fertile plains below, which they could cultivate with much less effort.

Farther south again, we come to the basin of the Bahr el-Ghazal (River of Gazelles) with its relatively even surface formed of sedimentary soil carried down by the floods from neighbouring higher ground through numerous shallow streams that can be followed only in their upper reaches. Their courses defy further identification, as they become lost in the midst of swamps long before they reach the river.

To the south-east the Sudan is bounded by the Ethiopian escarpments from which flow the Sobat, the Blue Nile, the Atbara, and their numerous tributaries, bringing with them vast quantities of silt which enrich the soil of the Nile Valley. Indeed, the Ethiopian plateau is an important source of Nile water for Egypt, and makes a priceless contribution to the economy of the Sudan.

THE SOIL

The study of agricultural development is closely interwoven with the type of soil available. According to J. Oliver, the soil of the Sudan can be divided into five main groups. In the extreme north soils are of loose sand with consequent high permeability; in the centre and west more 'stabilised sands' which allow for the growth of sparse savanna-type vegetation. The clay plain soils which extend from southern Darfur along the White and Blue Nile rivers to the Atbara (an area of about 200 kilometres wide) are the 'cracking soils'—alkaline, dark and heavy, which crack open during the dry season and, when not fissured, are highly impermeable; the evenness of the terrain is monotonous with only a variation where the swamps occur. The lack of gradient and partially embanked (levéed) rivers mean that the flooding is natural, and contribute to the formation of these swamps, known as the Sudd region. Higher and seasonal rainfall towards the south-west brings about a marked change in soil to tropical red loams which are highly permeable, while in the main southern upland areas soils are ferruginous; hence the name Ironstone Plateau for this region. Generally speaking, the soil in the Sudan changes gradually from north to south as the sand which covers the lands of the northern deserts and the northern parts of Kassala and Kordofan gives way to more fertile soil in northern Darfur and the greater part of Kordofan Province.

Farther to the south, the rich black soil, ideally suited for the cultivation of cotton, covers vast areas of southern Kordofan and extends in the east and south from Kassala to Equatoria Province. The tropical red loams in the Equatoria and Bahr el-Ghazal Provinces are extremely fertile. At present these lands are underdeveloped, though there is some cultivation of maize, groundnuts, sweet potatoes and bananas. American cotton, coffee, sugar-cane, rubber, palm-oil, cassava, pineapples and yams are also grown in many areas here.

THE LAND AND THE PEOPLE

Sudan stands out in sharp contrast to many of the Arab-speaking countries where between 35 and 40 per cent of the population are urban dwellers; only about 10 per cent of Sudanese live in towns. Khartoum, because of its central position in the richest zone, has the largest proportion.

In recent years the progress of agriculture in the Sudan has been notable. It is a land of vast potential providing sufficient water can be found. Tens of millions of hectares of arable land can be put under cultivation to develop an economy which rests on agriculture. Obviously this development depends upon the availability of water, for, with the exception of the regions irrigated from the Nile, the country suffers from an acute shortage. The importance of the regions which suffer from chronic drought lies in the fact that there are about 80 million hectares of land in them which could be made suitable for cultivation; in fact, the Sudan is bringing larger areas under cultivation, as can be seen from the following table (and with the development of water resources each year, more feddans are being cultivated):

92

Area (in feddans) under cultivation (1 feddan = 0.42 hectares)

	Irrigated	Under rainfall	Under flood
1969–70	1,774,621 (est)	7,900,000	150,000
1970–71	7,800,000	2,400,000	130,000

	Total cropped area
1969–70	9,820,743
1970–71	10,305,118

Source: Ali Ahmed Suliman, *The Middle East and North Africa 1975–76*, Europa Publications (22nd edition).

Note that while the areas under rainfall and under flood decreased, the irrigated area increased greatly, producing a total increase in the cropped area.

These regions already account for 70 per cent of the human and animal population. Furthermore, they produce about 80 per cent of the total world production of gum arabic. But because of the predominantly arid nature of these regions, herding remains the chief occupation of their inhabitants. Probably between 30 and 40 per cent of the population lead a nomadic or partially nomadic life in pursuit of pasture and water for their livestock.

Mostly throughout the Sudan settlement is directly related to water supply, which may be obtained by irrigation from the schemes (see map); by constructing *hafirs* as illustrated; or by sinking shallow wells such as those found in the Gezira Scheme—one of the most important irrigation projects since the construction of the three main barrages.

Because of the large percentage of nomadic peoples, one of the problems which the Sudan has to face is whether these peoples should be encouraged to settle and give up their practice of shifting agriculture, by giving them incentives such as mechanisation in the irrigated areas. With such large numbers and such wide areas, it is difficult to see how this could be carried out. The nomadic tribes do make a significant contribution to the economy of the Sudan in the form of meat production, milk, wool, and hides, but as stated on p. 97 if 'sedentarisation' were implemented, this would then mean that large tracts of land would simply lie fallow and ungrazed or uncultivated.

In the light of the varied nature of the land, the population of the Sudan may be divided occupationally as follows.

The Camel-herdsmen

Camels can live for several days without water, while cattle on the other hand can live for only a day, with sheep able to survive for two or three days. This has an effect on the areas of migration, camels being able to encompass a range of some 80 kilometres from water, but cattle much less—up to 20 kilometres. The pattern of the camel herdsmen's way of life in the territory known as El-Abala stretches both to the east and to the west of the Nile, south of the Egyptian border as far as the latitude of el-Obeid. Camels cannot venture farther south beyond this vast territory, since they could be fatally afflicted by the disease-carrying *ghaffar* flies. Large areas of the south, because of flooding of the plains, can practise agriculture only within the confines which these impose on farming; tsetse flies are a limiting factor in pastoralism in the south-west where the land is higher, while the increase in the insect pest population, such as locusts, mosquitos and grasshoppers, forces the migration of human beings and animals at certain times of the year. West of the Nile camels are found in the Sahara Desert as far as longitude 18°E across the borders of Libya and Chad: beyond this point the land is too arid even for the scanty requirements of these tough animals. North-east of the Nile, however, the herdsmen often drive their camels deep into Egypt's eastern desert where the highlands along the Red Sea give rise to at least some rainfall.

The Kababish are the best-known tribe among the camel-herdsmen west of the Nile. Their year is generally divided into two periods: the time of travel and migration, following the seasonal growth of grass, which lasts from June to December; and the time of settlement near the wells during the dry winter season. The latter is known as the *tadmir* season in the local dialect of these tribesmen. The absence of rain obliges them to settle with their camels and sheep and goats around a number of small oases and wells from January until April: they then resume their movement southwards at the beginning of the rainy season in early May. During this trek they often come into contact with sedentary tribes around Dar Hamar and Dar Hamid; but by the month of June, when the rain would have reached the north, they turn back in that direction, travelling in small groups. This journey northwards is called *el-noshouk*; the farther north they travel the scantier becomes the pasture and so they split into still smaller groups. Once the northern pastures can no longer sustain their herds, the Kababish people find their way gradually back again to the *tadmir* centres near the oases. Around these the pasture, untouched since January, sustains the animals until the beginning of the following year's rainy season.

East of the Nile live the camel-herdsmen of the Beja tribe. They lead very similar lives to those of the Kababish, although they are ethnically quite different. Whereas the Kababish are of Arab, Semitic origin, the Beja are of Hamite descent and still retain their original Tebdawi language as well as their own particular Arabic dialect. The Beja are mainly nomadic, but a subsection of their tribe, the Hadandwa, have taken to agriculture and the cultivation of cotton in a small way in the deltas of the Gash and the Baraka. Some of them work in the small goldmines around the town of Gebeit or in the docks of Port Sudan.

The Cattle-herdsmen

South of latitude 13° lies the territory of the cattle-herdsmen. Here the cow holds a very special place in the life of the inhabitants. It is the symbol of prestige and wealth, the unit of barter and trade and of all other transactions, such as the payment of dowries, ransom money and so on. It is the main ambition of every young man in the south to become a cattle-owner and to increase the number of his herd. The cattlemen lead the same sort of nomadic life as the camel-herdsmen. Few of them are sedentary, and even in the far south, where rain is plentiful and grass abundant, the inhabitants are usually on the move, because annual floods force them to leave their temporary dwellings.

Another interesting feature of the cattle nomads of the Sudan is that, although in most other countries flocks and herds actively seek water, here they actively avoid water during the flood season, and have to seek zones where the terrain is higher. These nomadic communities 'farm' an area which otherwise would be left fallow and unproductive. From recent figures, an approximation of the numbers of animals can be given thus: two million camels, eleven million head of cattle, seventeen million sheep and goats. These figures give some idea of the extent of the nomadic contribution to the Sudanese economy.

Although they derive their Arabic name *el-Baggara* from their occupation, the cattlemen are in fact composed of different ethnic groups. There are Arab tribes in the south of Darfur and Kordofan, and Nilotic tribes in the Upper Nile region— the Shilluk, the Dinka and the Nuer. These, in spite of their common origin and similarity of appearance, speak different languages and each tribe has its own customs and traditions. They practise agriculture in a small way, but cattle-raising is the mainstay of their economic life.

Their life is much easier than that of the Kababish tribes. Rain and pasture in their territory are abundant, but conditions constrain them to a semi-nomadic life of mobility and migration. During the dry winter season they live along the banks of the River el-Arab in the extreme southern corner of the provinces of Darfur and Kordofan; but with the coming of the rainy season the land becomes too marshy and their cattle fall victims to the disease-bearing tsetse flies. They then begin their trek northwards towards latitude 12°N where they spend the rest of the rainy season until the grass there begins to dry up, when they start their journey back to their winter homes in the area of the el-Arab River. It is here that they come into contact with the Dinka tribes who, in spite of their Arab origin, have been affected by this contact over the years and have taken on some of the physical characteristics of their Nilotic, negroid, central African neighbours.

At the beginning of the rainy season, and before starting out for the north, the Baggara tribes sow maize and millet which they harvest on their return at the end of the season. These crops, however, fall short of their requirements, so they have to supplement them with sugar, tea and cotton and by purchasing grain in the markets of Nyala, Abu-Zabad and other nearby towns, where they also sell their produce such as butter, cheese and hides.

Of the Baggara group, the Shilluk live on the Western bank of the White Nile and in the lower reaches of the River Sobat. They are gradually settling down into

The Roseires Dam in construction.

*River Dinder siphon: The supply is siphoned under the River Dinder. A weir controls the upstream level, while the downstream level remains at the corresponding discharge level for the canal.
Source: Sir M. MacDonald and Partners.

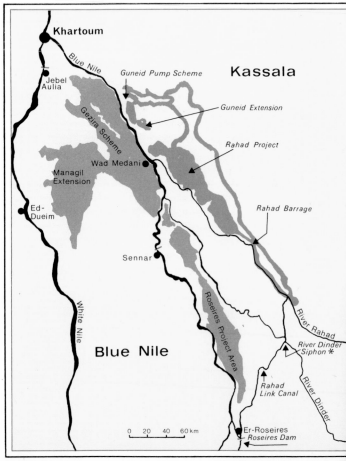

a more sedentary way of life, and depend less on herding than do the other Nilotic tribes, though they still own considerable numbers of cattle.

The Dinka and the Nuer, on the other hand, retain the simple and indeed primitive characteristics of the herdsmen's life. They occupy the flat regions in the lower basins of the rivers el-Ghazal, el-Jebel and Sobat. Their cattle are very dear to them and they are loath to diminish their herds either by sale or by slaughter. Their territory is vast and flat, criss-crossed by several minor tributaries of the Nile and covered with grass. They build their huts on what little high ground exists in this region, and to these points they move their cattle during the flood season. The lower plains, which the Dinka tribes call *tuak,* are their grazing grounds during the dry season. They never construct their dwellings on these plains; but they live on them as semi-nomads for about half the year, becoming sedentary cultivators of cereals on the higher ground during the other half. Their cattle is of the same large-humped variety as in the north, but the horns are longer.

The Sedentary Farmers

Between the two pastoral societies of the camel-herders and the cattlemen, in the north and the south of Sudan respectively, there live the sedentary farmers, whose territory includes both sides of the Nile Valley, the plateau of Kordofan and Darfur and the area southwards to the border between the Sudan and Zaire.

The Nile Valley to the north of Khartoum is only a narrow strip of fertile, alluvial land which is often broken by rock-ridges extending westwards, dividing the valley into closed, isolated basins and making communication difficult. It is generally a poor region; and for this reason most of the able-bodied men leave it to find work in other parts of the Sudan, and even travel as far afield as Egypt, leaving behind the old men, women and children. So the people of this area depend largely on remittances from members of their families who find work far from home.

The plateau of Darfur and Kordofan is inhabited by the Fur, the Nuba and other sedentary tribes who have taken up agriculture as their main occupation, and settled down to farming. So, too, have the Azande, who live further to the south on the higher land which forms the watershed between the rivers el-Ghazal and Congo. But the most economically important sedentary farmers of the Sudan are the inhabitants of the Gezira and adjacent irrigated areas (see map), the throbbing fertile heart of the country and the major centre of its economic life. For these are the people who cultivate the extensive areas of rich soil which lie between the Blue and the White Nile. Many aspects of the Sudanese economy depend on such schemes as the cultivation of the Gezira, as can be seen on p. 105.

IRRIGATION AND WATER RESOURCES

Water is the major factor which determines the population of each region of the Sudan as well as the main occupation. This applies not only to irrigation-water needed for agriculture, but also to drinking-water for men and animals alike, and to the water-courses used as means of transport and communication.

The agriculture of the Sudan depends both on irrigation and on rain. With the exception of the Red Sea coastlands, rain falls in the Sudan in the summer. Moderate and evenly-distributed rainfall helps to increase production; whereas excessive prolonged rains are as damaging as drought.

The central Sudan is the main region of xerophilous agriculture (that is, the cultivation of crops that flourish in dry climates); but the rainfall in this area is rather erratic. As we progress farther south the fluctuation of rainfall from year to year becomes less noticeable.

Irrigation methods vary from one region to another. Basin irrigation is predominant in the Northern Province, irrigation by inundation is common practice in the Deltas of the Gash and the Baraka, canal irrigation is used in the Gezira, and irrigation by pumping takes place alongside the White Nile and the main river north of Khartoum. The ancient methods of irrigation such as the *sakia* (water-wheel) and the *shadouf* are also still practised.

In the western parts of the Sudan agriculture is confined to various localities in Darfur Province, and here small dams of a temporary nature are constructed to retain water supplies, or *hafirs* made—shallow basins which will retain rainwater often throughout the rainless period. Work is going on in the use of polythene for their construction, as shown in the photographs of experimental *hafirs*, and further water retention methods are taking shape in the form of excavated underground tanks, into which rainwater infiltrates from a small catchment area on the surface.

Basin Irrigation

Basin irrigation allows some agriculture to be practised in the north where rain is either scarce or non-existent. The system is the same as in Egypt; but the areas thus irrigated in the Sudan constitute small pockets in the desert adjacent to the Nile north of Khartoum, with a total area of only about 30,000 hectares. Even this small area is rarely wholly cultivated in any one year.

Irrigation by Inundation

This is the method used in the deltas of Kassala and Tokar into which flow the rivers Gash and Baraka respectively. Flash irrigation is highly developed in the valleys of the Baraka and Gash rivers, run on the same lines as that in southern Yemen. The principle is to divert outwash streams from the plateaux by rotation over large areas where the streams emerge on to more level plains. Both rivers have their sources in the Ethiopian-Eritrean highlands and run down into the east of the Sudan where they fan out to form two fertile deltas, the first in northern Kassala and the second near Tokar. The two rivers rise in summer and the flood-water reaches the deltas between 20 June and 10 July. The flood-season continues for about three months, on average, and the area of land which is consequently put under cultivation depends on the size and duration of the floods each year. Until very recently, the two rivers ran their courses at random, but a barrage has now controlled the waters of the Gash and allowed the arable land to be divided into basins fed by irrigation canals. The average area of cotton cultivation in the Gash Delta is estimated at about 15,000 hectares. There is a plan to build a dam on the Gash, but no steps have yet been taken to implement the project, though a dam has been built on its course at Tessenei over the border in Eritrea.

The Baraka differs from the Gash in significant ways. The Baraka's normal flood-water amounts to no more than 20 or 25 per cent of that of the Gash and there is no way in which to predict or control the distribution of this flood-water. Thus cultivation in the Delta of Tokar is somewhat risky; for, when the farmer has sown the seed, he can never tell whether the land will receive enough water during the flood season to grow a good crop; or indeed, whether he will have any crop at all. On the other hand, it is practically impossible to regulate irrigation in this region through canalisation and dividing the land into separate basins as in Kassala, because of the frequency of fierce sandstorms in the dry season. These would clog the canals and it would be necessary to clear them every year. The *haboob* is particularly associated with the Sudan—a violent sand-wind of dramatic proportions forming a wall of dust several thousand metres high and several kilometres long. It sweeps over the arid areas where there is little vegetation, generally around May and June, the driest time.

Canal Irrigation

One of the most important irrigation projects in the Sudan has been the construction of the Sennar Dam on the Blue Nile. The potentialities of the Gezira as an area for irrigation were first recognised in 1845; and by 1913 all the preliminary studies and surveys for the construction of the dam were complete.

The construction of the dam began in 1914, but the outbreak of World War I interrupted further progress on the scheme. Work was resumed in 1922 and the dam, together with the construction of the irrigation scheme related to it, was completed in 1925. The dam is nearly 3 kilometres long, with both piers (835 metres on the eastern side and 582 metres on the western side) built with solid rock-fill, while the middle section—1,607 metres long—is made of granite. This section has 80 principal gates, each of which is $8\frac{1}{2}$ metres high and 2 metres wide, with auxiliary gates between and above the main ones.

The Sennar Dam impounded initially 636 million cubic metres of water, but in 1956 this was increased by raising the water level of the reservoir. This structure feeds water by means of regulators to the Gezira Plain, which is an area of some 1,200,000 hectares lying in the triangle formed by the Blue and White Niles as they come together to meet at Khartoum. Here the irrigation scheme which was under

An experimental *hafir* fenced off to avoid damage by cattle. It is lined with two layers of polythene. The ground beneath the lower membrane is treated with insecticide, as is the clay slurry between the membranes.

Testing the decrease in volume of water.

Termite damage to the exposed edge of the membrane lining.

consideration as far back as 1845 has produced the main cotton-growing area of the country.

The Rahad Project area, although more distant from the Blue Nile at Roseires, has a number of advantages which influenced its choice for development: open grassland (no timber to be cleared); uniformity of soil (the greater part of the area is covered by dark-coloured cracking clays similar to those found in the fertile Gezira area). The clay soils are mostly non-saline and moderately alkaline—the penetration of water in clay soils, with their low permeability, depends to a great extent on the degree of cracking. The land is also regular, and excellent for a gravity canal system, with a gentle slope towards the River Rahad. The area has some population, an important consideration when the other project areas are almost uninhabitable and are situated at great distances from towns.

The cropping pattern for the type of clay soil which would allow for the maximum return on both capital and land is: one-third medium staple cotton; one-third groundnuts; one-sixth dura (a grain sorghum); and one-sixth fallow. Dura under irrigation is not a profitable crop, though indigenous to the farmer's way of life in the Sudan. Alternative crops which have a similar growth pattern could be introduced if the farmer could be discouraged from growing dura and encouraged to buy it from large mechanised schemes close by.

Rotational patterns are based on the fact that water for irrigation has to be supplied from November to June through the 191-kilometre link canal from the Blue Nile at Roseires to the River Rahad. Economy at peak demand periods has meant a smaller link canal. This rotational pattern has been formulated to make the best possible use of the flooding period of the Rahad River between August and October. Thus maximum economy is coupled with maximum output. Ultimately the Rahad Scheme will take in the Guneid extension. (See map p. 96.)

These projects use water stored in the recently completed Roseires Dam on the Blue Nile, made available to the Sudan under the 1959 agreement. The share of the Nile's water prior to this, during the restricted period from January to July, was obtained from the 781 million cubic metre contents of the Sennar reservoir. Sudan's share under the new pact was boosted to 18·5 milliard cubic metres yearly.

Pump Irrigation

Pump irrigation is practised mainly on the Nile over an area of about 150,000 hectares, containing some 1,400 farms which range in size from a few to several thousand hectares. The larger farms, about seventeen in number, are cultivated by tenants and owned by the Government. Ten of these are located in the Northern Province where they were initially set up in an effort to ensure that the inhabitants of this region were protected against recurring famine. For this reason, these farms were in the beginning given over entirely to the cultivation of dura, but when the danger of famine receded after the successful completion of the Gezira scheme, the pump-irrigation schemes were turned towards the cultivation of cotton as well. There are, too, a number of co-operative pump-irrigation schemes in the north. Here, all the members first take half their crops; working expenses, cost of maintenance, interest and payment of capital are then deducted from the proceeds of the other half, and the remainder is shared out. There are also smaller farms worked by individual owners; but generally speaking fully mechanised irrigation schemes are still in their infancy, and pumps are used side by side with the primitive and traditional methods of the *sakia* and the *shadouf*.

Judging by the results that have so far been achieved, however, it may be said that pump-irrigation has been successful and has given encouraging results, both on the government's and on the private farms, in the more northerly regions of Wadi Halfa and Dunqul where the irrigated land has produced good crops of American cotton, wheat, maize, millet and vegetables. In the regions of Berber and Shendi farther south, on the other hand, pump-irrigation schemes have been less successful owing to heavy infestation by pests and parasites.

The Sakia and the Shadouf

Ancient water-lifting appliances, mainly the *sakia* (water-wheel) and the *shadouf*, are still much used in the Sudan. In fact these are the most important methods of irrigation used for the smaller plots of land adjacent to the Nile in the north. The *sakia* is more commonly used in the Northern Province, where more than 90 per cent of the total number of these old contrivances are to be found; whereas the *shadouf* is more often seen on the banks of the Nile farther south.

Small patches of land are often cultivated with the aid of these antique methods of irrigation to produce several crops each year; thus the soil becomes exhausted and in addition the underground water-table may be dangerously raised. Fortunately, however, the flood-water revitalises the soil with its silt deposits.

Similar in design to those used in Egypt in the former practice of basin irrigation are the earth embankments known as *terus* which are built on interfluves to retain flood-water. (Evaporation and seepage losses can account for 50 per cent of the whole.) While the traditional use of *shadouf* and *sakia* is still employed near to rivers, there is an ever-increasing trend towards mechanisation, and motor pumps are common along the banks of the main Nile and both Blue and White Niles. The height of lift is limited to 19 metres.

Abger pumping station, one of the many similar stations on which cultivation depends.

THE LAND AND THE WATER

Of the total area of the Sudan, which amounts to about 250 million hectares, about 40 million hectares are arable but as yet uncultivated. The area of the cultivated land is about seven million hectares, and the annual crop area is estimated at around four million hectares, of which three million rely on the rainfall and only one million on irrigation.

The drinking-water requirements of man and animal in the rural areas of the Sudan are estimated at about 240 million cubic metres per annum. Water is, as we have seen, plentiful during the rainy season which lasts for about three months; so it is necessary to conserve or provide some 190 million cubic metres of water to tide over the country for the rest of the year and avoid the danger of thirst. For this reason, a special department has been set up in the Sudan with the sole object of developing underground water-resources. Since its creation in 1967 this department has succeeded in augmenting the drinking-water supply in the countryside by about 30 million cubic metres per annum.

Water for cultivation is, however, another matter. With irrigation a vital necessity for agriculture both in Egypt and the Sudan, the flow of water developed in the Nile system is not sufficient for present daily needs. According to H. E. Hurst, 'while on average 84,000 million cubic metres of water enter Egypt at Aswan, this figure can fluctuate between one half to almost twice as much as the mean quantity.' An agreement between Egypt and the Sudan in 1959, whilst making better provision for the Sudanese requirements, encompassed the proposals for the Aswan High Dam. The Sudan has now lost an area of cultivation in the Sudanese part of the Nile from Wadi Halfa northwards. And there is in addition a great loss of water as the Nile flows through the Sudan: by evaporation, by percolation into the substrata, and by transpiration of sudd-type vegetation. Another problem is that often flooding of the rivers occurs during the rainy season, when the water is in excess of what is needed at that time.

Forest and animal products, as well as aromatic crops, constituted the mainstay of the Sudanese economy for a very long time. After the First World War, however, agriculture began to play an ever-increasing role which eventually changed the entire structure of the Sudanese economy, to the extent that it is now mainly dependent on agriculture, as we have already seen. Inevitably this transformation has had a marked effect on the social and economic life of many of the Sudanese people; since, with the progressive expansion of agriculture, the population tended to settle down on the land and to lead an increasingly sedentary life. This trend has been accompanied by a corresponding expansion in the Government services dealing with water-conservation, irrigation control, regulation of the agricultural cycle, experimental farming and research, soil conservation and pest control.

As it is at present practised, agriculture in the Sudan may be classified into two categories: shifting and sedentary agriculture. Let us examine these two types in a little more detail:

Shifting Agriculture

This form of agriculture is practised in most of the plains of the Sudan where the rainfall is sufficient to allow for the cultivation of the land. The human effort required for this form of agriculture is indeed minimal, since it is confined to clearing the surface of the land and tilling it with simple, often primitive, implements. No fixed agricultural cycle is observed and the land receives no fertiliser; when it begins to show signs of exhaustion, the tiller simply leaves it to go and find another suitable virgin plot where he repeats the same process all over again. In view of the vastness of the area of land available and the comparatively small population, this form of shifting agriculture has so far presented no acute problems and there are as yet no compelling reasons why the population should change their habits in this respect.

This form of agriculture is practised in several parts of the country, notably in the gum-arabic areas in the provinces of Kordofan and Darfur, on the terraced mountain-slopes, and in the tropical regions of the south.

A related type of agriculture, which is also common in the Sudan, is the *hariq* (fire) method of cultivation. The difference here is that the dried plants left from the previous year are burnt on the land before the new seeds are sown. After the same plot has been tilled for two successive years it is usually left fallow for the same length of time, thus allowing it to acquire a cover of tall thick grass. So, just before the next rainy season begins, the old grass is fired and the land made ready for the new crop. It is generally believed that the soil gains in fertility by this practice, but the question is a highly controversial one, and there is no doubt that it can contribute to soil erosion.

The Nilotic and other tribes of the south practise shifting agriculture with variations. The Shilluk cultivate the land around their villages near the river banks for three successive years; they then leave it to move to other plots a little farther away. The Nuer, on the other hand, cultivate the same plot year after year until it is exhausted and then move their entire village to a new area.

The Dinka do not cut down the trees which they find on the land but grow their crops beneath and among them. They cultivate the same plot for a period of not less than five years; and these cultivated plots are often several miles away from their villages. They grow millet on the open land and sow sesame seeds underneath the trees.

Generally speaking, it can be said that there is no problem of land shortage in the Sudan. But with the progressive increase in the population, the raising of their living standards and the introduction of modern machinery into agriculture, an increase in the demand for agricultural land must eventually follow. In fact, some regions are already somewhat over-populated; especially those where water is scarce. Elsewhere, there has been a deliberate gathering together of the population in an endeavour to seek protection from sleeping-sickness (*trypanosomiasis*), which is carried by the tsetse fly. In such areas, the period during which the land is left fallow is inevitably shortened and indications of soil exhaustion are becoming more and more apparent; so much so that the authorities have already taken steps to provide new land for the resettlement of the peoples concerned.

Shifting agriculture, especially the method of firing, has been responsible for the total destruction of large tracts of forest in the south. Realising the consequent damage to the country's economy, the Government developed a scheme for forest protection and prohibited the practice of shifting agriculture in the forests.

In short, although this form of agriculture is not yet seriously damaging, it does have its own peculiar problems which have to be tackled, and the experiment of the Zande Scheme, which we will describe below, has shown what the Government can do to help.

Sedentary Agriculture

This is indeed the most important form of agriculture in the Sudan, in terms of production; but the area in which it is practised represents a mere fraction of the immense expanses of the Sudan's arable plains. Sedentary agriculture is practised in separate areas situated in different regions, and it depends on various systems of irrigation. The main areas under this form of cultivation are the following.

The Gezira

The Gezira Scheme is perhaps the most important agricultural project on the Nile, after the High Dam. Gezira means 'island' in Arabic and the name refers to the great plain of about two million hectares which lies between the Blue and the White Nile before they meet at Khartoum. The possibility of irrigating the Gezira was first given serious consideration early in this century, when several experiments were begun to cultivate long-staple cotton on several farms of some thousands of hectares each and watered by pumps from the Nile. The encouraging results induced the Government of the Sudan to push ahead with the major scheme in co-operation with the Sudan Plantation Syndicate. It was agreed that the Government would build the Sennar Dam and the main canal system, while the Syndicate would be responsible for agricultural and business operations.

When the Sennar Dam was completed in 1925, the Gezira Scheme was first begun on an area of 100,000 hectares, of which 33,000 were devoted to cotton. Gradually the irrigated area was increased to about 400,000 hectares in 1958, when the Managil Scheme (an extension of the Gezira) was also inaugurated to add a further 300,000 hectares to the total irrigated area.

Intensified cropping in the Gezira Scheme has been boosted by water from the Roseires Dam; new cash crops have been introduced such as groundnuts and wheat; in 1976 the area of the former crop was increased from 370,000 to 600,000 feddans, and the area of the latter from 400,000 to 600,000 feddans. This has altered the pattern of organisation, and has had repercussions on the relationship between the cultivator and authority throughout the Sudan. Already there is a Trade Union among cultivators in both Government and private irrigation schemes.

The Government did not expropriate the land for the Gezira Scheme but leased the land within the irrigation area for an annual rent of 10 piastres (ten pence), or, where land was needed for canals or other similar works, compulsorily purchased it.

The British-owned Sudan Plantation Syndicate and Kassala Cotton Company continued to administer and finance the scheme, the land for which was provided by the Sudan. This tripartite arrangement continued until the end of the concession in 1950, whereupon the scheme, though not the land, was nationalised and the Gezira Board was created to manage it. The Gezira Board was established on the pattern of the Tennessee Valley Authority in the United States, and has maintained some degree of independence from direct Government control. The method of husbandry and the share cropping system continued along the same lines, though there have been extensions in area, and some reduction in the rotation of cotton in order to reduce the incidence of disease.

The Gezira Scheme was first established on the basis of sharing the net revenue from the cotton crop between the Government, the Syndicate and the tenants at the rate of 40, 20, and 40 per cent respectively. After the nationalisation of the scheme, however, these proportions were revised as follows:

42 per cent to the Government
42 per cent to the tenants
10 per cent to the Gezira Board
2 per cent to the social services in the Gezira area
2 per cent to form a special reserve fund for the tenants
2 per cent to local government councils in the Gezira.

For many years now, the tenants have made various demands, including the raising of their share of the revenue to 50 per cent and participation in the management of the scheme.

This system of share cropping is similar in pattern to that long followed along the Nile, whereby the cultivators provide their own labour from their families or on hire, while land, water-wheels, animals for tilling, or mechanisation (e.g. tractors), are obtained from outside sources. Nevertheless the Gezira has some 'arrangement' of its own: the farmer is a tenant given an annual tenancy by the Government who in turn lease the land from the original owners; his water is free for his own food crops and fodder, but in all other ways there is strict supervision of land, water, and the management of the cotton crop.

The scheme is worked in 'blocks', each block averaging 8,000 hectares. The block inspector, who has a small staff, is responsible for both the supervision of the cotton crop and water control. He prepares weekly schedules of the water requirements, which are passed on to the Ministry of Irrigation, who regulate the flow from the Sennar Dam accordingly.

The Managil Scheme

After independence in 1956, there was an increase in the expansion of agricultural production by means of irrigation, the Managil Project being one of the major developments, and in fact an extension of the Gezira Scheme. It doubled the total area covered by that scheme and, because of more intensive rotation, more than doubled the cotton area. The Managil Scheme was begun in 1957, when a new dam was built at Roseires on the Blue Nile, and was completed in 1962, being based on the principles underlying the Gezira Scheme, except that the Government, which already owned one-third of the land, had bought up the remaining area from its private owners. Another difference was a reversion to the original three-course crop rotation, instead of the four-course rotation adopted in the Gezira, and the establishment of the standard unit for tenancy at 15 feddans instead of 40 feddans as in the Gezira. Thus a greater number of tenants could be included in the scheme and its benefits spread more widely. The smaller unit was also designed to encourage the tenant and his family to work the land by themselves, without needing help from hired labourers.

The 40 feddans of each tenancy in the Gezira meant that four fields were allocated to each. The fields are separate from each other so that a pattern of rotation can be practised: in one year one field grows cotton, two are lying fallow, one grows dura or is even resting as well. According to Barbour, 'the initial allocation followed the settlement (in a legal sense) and registration of almost the whole of the area between the two Niles north of Sennar. Persons with more than 40 feddans were granted tenancies as of right. Those with more than 20 feddans were held to have preferential tenancies.' However, as the number of holdings at the beginning exceeded those who wanted them, tenancies were granted to many non-Sudanese westerners in order to get the ground cultivated.

Another difference between the Gezira and the Managil schemes, at the outset, was the fact that Gezira began as a commercial scheme, with social services incorporated into it later; whereas the Managil extension was planned from the beginning to cater for social services side by side with its commercial aspects. Its villages were thus established on the basis of well worked-out plans which also allowed for future expansion.

The Gash Delta Scheme

During the initial stages of the Gezira scheme, experiments were being made with cotton-growing in the Gash and Tokar Deltas, where a simple system of regulators and canals allowed for the cultivation of about 25,000 hectares each year by flash irrigation. The main problem hampering this scheme had been the lack of communications with the coast; however, this problem was overcome in 1924 when the main town of the Delta, Kassala, was connected by rail with Port Sudan. This was part of the agreement reached between the Government and the Kassala Cotton Company in 1922 for the development of cotton-growing in this region. In 1928 the Government took over the whole project and created a special board to manage it, on the same lines as the Gezira scheme.

The areas cultivated in these deltas vary annually—40,000 feddans in the Gash and 80,000 feddans in Tokar can be considered 'typical'. Mainly long-staple cotton is grown, with areas of poorer soil around the edges of the Gash devoted to the cultivation of sorghum. The Gash Delta land is let to about 7,000 tenants whose holdings vary in size between 2 and 20 hectares. The land is distributed on a tribal basis, in that tribal chiefs or their agents are each allotted an area of land which they then redistribute among the members of their tribe. Cotton is cultivated in those parts of the Delta which receive sufficient water, according to the state of the flood each year and the difficulties of water control. We have seen already how erratic and uncertain this is.

The Gash Board has a complex system of water control, and is responsible for the supervision and auction of the cotton crop, although the tenants are charged for their cotton seed. In Tokar, the Government distributes the seed and runs loan schemes.

Replacing cotton with castor in the Gash, according to Tewfiq Hashim Ahmed, has failed to achieve increased income sufficiently to meet even the costs of administration. 'Although most of the people who take part in these schemes can be classified as semi-nomadic, recent estimates do not show that cattle are a source of wealth to more than a handful.'

The Baraka (Tokar) Delta Scheme

The Tokar Delta was to begin with a silt plain formed by a small inland delta approximately 160,000 hectares in area, on to which poured an annual flood carried down by the River Baraka from the Eritrean highlands. This was the first area of the Sudan into which long-staple cotton cultivation was introduced, though since the Second World War more water has been used for food crops. Hence the Tokar Delta Scheme was indeed the forerunner of all the agricultural development schemes initiated by the Government of the Sudan. In the absence of any certain rights of ownership, the entire area of the Delta was declared to be Government land, and allocations made annually by the Government to applicants who wished to cultivate in accordance with the guarantee system, by which the land was made over to the heads of tribes who were then made responsible for satisfying the demands of the members of their tribes as they saw fit. The allocations per tenant are smaller than in the Gezira, and the income correspondingly less. From 1943, however, the Government, having recognised the inadequacy of this system, has embarked upon a new policy of gradually decreasing the areas allotted to the tribal chiefs and making more allocations directly to individual cultivators in the hope and expectation that the guarantee system will ultimately die a natural death. The Government receives 20 per cent of the crop in rent and taxes.

Pumping Schemes on the White Nile

The construction of the Jebel Aulia Dam led to the inundation of a vast tract of land which was previously inhabited by about 7,000 semi-nomadic families. It was necessary, therefore, to find these people another means of livelihood. A large part of the £3 million which the Egyptian Government paid to the Sudan in compensation for the water from the Jebel Aulia was used to develop new agricultural land for the benefit of immigrants from the inundated regions. The pattern of administration is similar to that of the Gezira, with a slight difference in the sharing system of the cotton crop. These schemes, which depended on pump-irrigation, were eventually implemented in the selected regions of Abd el-Magid, Hashaia, Umm Garr, Wad Nimr, Ed-Dueim, Uakra and Abger.

The Zande Scheme

Towards the end of 1943, the Sudan Government embarked upon an important economic undertaking in the territory of the Zande in the extreme south. The object of this experiment was to render these remote parts of the Sudan, so far away from the coast and its ports, as self-sufficient as possible. The idea was to develop some kind of production and some relevant local industries so as to induce the Zande people to settle down and thereby to promote their social development and advancement.

Cotton formed the core of both the agricultural and the industrial experiments. Trials were made, and are still in progress, of the cultivation of cotton in several parts of the Zande area. They have all proved successful so far; but the main areas of cotton cultivation in this region are still centred around Maridi and Yambio on the Sudan–Zaire border.

Opinions differ about the degree of success achieved; nevertheless its importance lies in the fact that it pointed the way to the exploitation of the hitherto neglected regions in the south, in the interest of the local inhabitants and of the Sudanese economy as a whole.

Other schemes

Other, lesser schemes include the Abu Jebel scheme in Kordofan which was begun in 1945 following the construction of a special dam on the river to store its water and to regulate the irrigation of the more fertile areas by means of a $4\frac{1}{2}$-mile long irrigation canal. The crops are sown in the rainy season, but they are irrigated later from stored water. The area of cotton cultivation in this scheme is now estimated at about 7,000 hectares. There are also many privately-owned schemes depending on pump-irrigation; these have increased in number in recent years.

Two sugar schemes each of about 30,000 feddans and designed to produce 60,000 tons of sugar annually are of particular interest in view of the world shortage of this commodity. One is based on pump-irrigation, while the other is attached to the Khashm el-Girba development on the Atbara River. In 1974 moves were made to increase the production of sugar; 100,000 feddans of land near Kosti should be producing 350,000 tons of white sugar annually by 1979.

109

The Agrarian Structure of Four Classes of Tenure in the Sudan

	South	Northern Province	Gedaref District	Gezira Scheme
Locality :				
Type of Tenure :	Communal	Individual ownership (Family farm)	Annual leasehold	Triple partnership
Organisational Structure :	'Traditional' Managerial skill Communal labour Family labour: male, female, child	Head of family Family labour	Licensee or hired occupier, usually with technical experience Hired labour Skilled labour Migratory labour	Sudan Gezira Board and tenant Controlled labour (industrial) Contractual labour Tenant's family and hired labour
Type of Farming :	Intensive cropping and grazing Dry farming Non-mechanised Diversified production Subsistence	Intensive farming Irrigation: flood, basin, pump, water-wheels and buckets Non-mechanised Diversified production Subsistence, internal market and export	Extensive farming Dry Mechanised Specialised Internal market predominantly	Intensive farming Gravity-fed irrigation Mechanised and non-mechanised Specialised Primary export, subsistence and internal market
Size and Operation :	A basic common unit in terms of village family heads Carrying-capacity of range as a measure of size Land worked by household	Small holdings Fragmented holdings (0·5 to 1·5 hectares) Worked by owner	Rent system (fixed rent)	Share system

110

Source: Farah Hassan Adam, 'Contribution of Land Tenure Structure to Agricultural Development in Sudan through Incentives, Knowledge, and Capital', *Research Bulletin* No. 5, 1966, University of Khartoum.

LAND TENURE

Land tenure in the Sudan is governed by the Titles to Lands Laws of 1925 which replaced all previous legislation. All land that had not been registered in the names of individuals was considered to be Government property, though it was left as areas of concession in the hands of the individuals or tribes who habitually lived on it, without, however, also conceding them the right of legal ownership.

Naturally, in a society which was being gradually transformed from the nomadic to the sedentary state, the land-tenure system had to go through several stages. (See table.) But, since the process of transformation did not follow the same pattern everywhere in the Sudan, the various stages of the land-tenure system all existed at one and the same time. Tribal ownership, which was peculiar to the nomadic population, figured very largely over much of the country. There was also village-ownership, whereby the land was the common property of all the sedentary inhabitants of the village; this prevailed in some parts of the country, especially in Kordofan Province; and there was private ownership, which was more common in the Gezira and the Nile Valley in the Northern Province. Although the Government recognised the latter kind of ownership, the 1935 Titles to Land Law prohibited the owners from disposing of their lands without the Government's prior approval.

This measure was intended to prevent speculators from taking advantage of small landowners to acquire large holdings. The same law stipulated that all forest areas and uncultivated lands were Government property unless it was proven otherwise. It also stipulated that no right of ownership arose from the intermittent cultivation of any plot of land.

Government lands are thus classified into two categories:
1. Lands free of concessions such as those of the Tokar and Kassala Deltas, where the Government refused to recognise any rights of private ownership, though it concedes the right of certain individuals and tribes to have the first option to cultivate particular areas under the Government-controlled schemes.
2. Government lands proper, which include those on which nomadic or semi-nomadic tribes live in the provinces of Kassala, Kordofan and Darfur. Each tribe has its own grazing territory, known locally as its *Dar*—the Arabic word for 'home'.

In southern Sudan, however, no attention has ever been given to the registration of titles since the area of land available is so vast as to make this question completely irrelevant. Land is generally shared by all members of a tribe, although technically it remains the property of the Government.

ANIMAL HUSBANDRY

Herding in the Sudan dates back to at least the third millennium BC. To this day animal wealth comes third in importance only to the production of cotton and dura.

Leaving out of account some animal husbandry practised in the Government schemes of the Gezira, the Northern Province and on the Blue and White Niles, the animal wealth in the Sudan, estimated at about 24 million head, depends almost entirely on natural pasture. Research is being conducted to determine how grazing could be used to restore the fertility of the land damaged by shifting agriculture.

The area of forests and pastureland in the Sudan has been estimated at about 61·3 per cent of the total area of the country; but in certain parts camels and goats feed on trees and shrubs, as is the case in north-western Kassala and to the east of the Blue Nile. Wild animals, of course, also feed mostly on trees and shrubs.

The Sudan is perhaps the only Arab country which enjoys a surplus of animal wealth. The United Nations Food and Agriculture Organisation forecast an increase in the Sudan's production of beef and mutton of between one and two per cent per annum; but the actual increase achieved between 1948 and 1963 was about 18 per cent, which indicates the rate of expansion in livestock production, and the revenue return is far greater than that of cotton in an economy which is dominated by agriculture. The Sudan is a progressively developing country in the Arab World.

The geometry of an irrigated depression.

Work on the construction of a gravity-flow canal.

REFERENCES

K. M. Barbour, *The Republic of the Sudan: A Regional Geography*, University of London Press, 1961

Arthur Gaitskell, *Gezira: A Story of Development in the Sudan*, Faber, London 1959

Sir M. MacDonald and Partners, *Water Development Survey Project in Darfur Province, Sudan*, London and Cambridge 1968

G. F. Albany Ward (Sir M. MacDonald and Partners), *Sudan Pre-investment Surveys*, London 1968

International Development Association, *Rahad Irrigation*, Washington 1973

W. B. Fisher, *The Middle East*, 6th edition, Methuen & Company, London 1971

The Middle East and North Africa 1975-76. A Survey and Reference Book, 22nd edition, Europa Publications

Gezira Board Annual Report, Khartoum 1972

J. H. C. Lebon, *Land Use in Sudan*, Geographical Publications, Bude, England 1965

Marion Clawson, Hans H. Landsberg, Lyle T. Alexander, *The Agricultural Potential of the Middle East. Economic and Political Problems and Prospects*, Elsevier, New York, London, Amsterdam 1971

D. S. Thornton, *Contrasting Policies in Irrigation Development, Sudan and India*, University of Reading, 1966

National Academy of Sciences, *More Water for Arid Lands. Promising Technologies and Research Opportunities*, Washington 1974

113

The Libyan Arab Republic

Prior to the formation of a modern state in 1951, Libya was little more than a scattered entity, acting as a route from east to west and as a buffer zone between the Mediterranean Sea and the continental Sahara. This huge Arab country, of 1,759,540 square kilometres, lying to the west of Egypt and the Sudan, with Chad and Niger to the south, Algeria and Tunisia to the west, and the Mediterranean Sea to the north, has a small population of 2,200,000 people who, against a background of a disturbed history of foreign domination, are in the process of economic transition brought about to a great extent by the exploitation of petroleum.

The size of the population in relation to so vast a territory, a large proportion of which consists of desert, and the fact that three-quarters of its peoples are concentrated in an area within 20 miles of the sea, constitute two factors of vital significance in the development of the country as a whole.

Geographically, Libya is made up of three separate regions, each with its own distinctive characteristics which are more allied to the countries neighbouring them than to each other: Western (Tripolitania), Eastern (Cyrenaica), and Southern (Fezzan). In 1963 these provinces were subdivided into ten *Muqatta* (regions).

Since oil was first discovered in Southern Libya (Fezzan) in 1957, Libya has become one of the world's five largest oil producers and exporters. Before then, agriculture was of major importance, employing 70 per cent of the labour force; a dramatic decline to 40 per cent today emphasises the need for the priority the Government has given to expansion. (The First Plan covered the years 1963–68, and a threefold increase in revenue was budgeted for in the Second Plan, which ended in 1974.) Much of the money expended was given for such important aspects as fertilisers, farm machinery, agricultural credit and price support, with a consequent increase in output. Nevertheless the hard facts remain in the scarcity of rainfall, the shortage of water, and land which for centuries has been overgrazed and become increasingly impoverished. These are great obstacles to expanding agricultural development, and stand out in sharp contrast to the oil boom.

The report of the First Five-Year Plan said that although a general improvement in the standard of living had been attained, agriculture would continue to be the main source of employment and the utmost attention would be paid to this. Its development would therefore be the most important method by which the majority of the people could benefit from the wealth derived from oil.

The tiny population, scattered as it is over a vast territory, presents enormous problems of supply and production, the presence or absence of water being really the determining factor in the settlement of three-quarters of the population. Therefore the two major cities of Tripoli and Benghazi are a great urban attraction for many, as is the completely new Federal capital Beida in Eastern Libya. The harsh nature of the land and the climate has, as elsewhere in the Arab World, inevitably led historically to a high nomadic and semi-nomadic population. According to the 1964 population census, 17 per cent of the population were classified as nomadic and semi-nomadic, with a significant proportion of true nomads migrating in orderly patterns or pathways throughout Libya, while the majority of semi-nomads follow a seasonal migration from the sedentary areas such as the two northern Jebel zones. As with Egypt and the Sudan, there are two major facts about Libya. One of these, the desert, is common to its eastern neighbours; the other, oil, is not. The almost totally Saharan environment of the country made it scarcely viable economically until recently; the exploitation of its natural oil wealth in the last fifteen or so years has raised the standard of living for many, particularly urban dwellers, but at the same time has brought with it social problems analogous to those of the very poor man who suddenly inherits a vast fortune.

PHYSICAL FEATURES

The country can be likened to a mosaic pattern, territory which is settled (because better watered) being dotted among vast expanses of arid and semi-arid land.

Climatically Libya is very similar in pattern to Egypt, with the slight difference that rainfall in some zones is heavier than that encountered in Egypt; in geographical formation it also resembles Egypt more closely than it does Morocco or Algeria.

Western Libya (formerly Tripolitania)

Tripolitania, or the Western Region as the Libyans call it, can be likened physically to three broad steps rising generally from north to south. In the far north there is a triangular-shaped coastal plain, known as the Jefara, some 300–400 kilometres from east to west and about 150 kilometres at its widest, which contains a relatively small strip of Mediterranean-type vegetation. Here Tripoli, the capital, is situated, with a generous—by Libyan standards—rainfall of 400 mm per annum. A few miles inland lie a series of massifs known collectively as the Jebel, which has a cover of dry Mediterranean brush. Although there is less than 150 mm of rainfall per annum, there are scattered pockets of agriculture which depend on what is known as controlled perennial irrigation from springs and wells, and on 'wadi culture'—uncontrolled irrigation by flood. These methods of water use can be found mainly on the slopes of Jebel Nafusa. Olives and figs are grown here. Beyond the Jebel to the south there lies an upland *hamad*, or desert plateau of sand, stone and scrub, 400–500 metres above sea level, which stretches south for several hundred miles until it reaches first southern Libya (Fezzan) and ultimately the Tibesti Mountains of the Central Sahara.

According to Lalevic (1967), the majority of the 105,000 hectares of irrigated land in the north depends on ground-water accumulating in the ancient sediments of the Jefara coastal plain which have, as Hill (1960) states, formed two water-tables. This water around Tripoli can be reached by using the *dalu* (leather bucket)

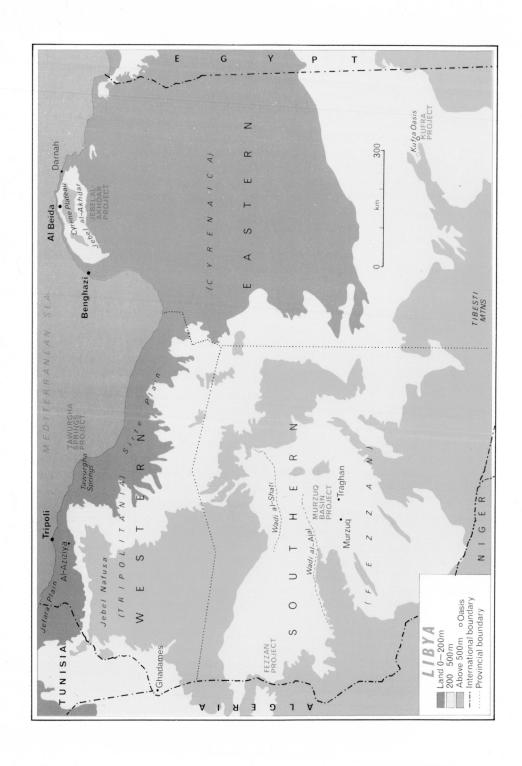

LIBYA

Land 0–200m   o Oasis
200 500m
Above 500m   —··— International boundary
              ·········· Provincial boundary

EGYPT

EASTERN

(CYRENAICA)

Kufra Oasis
KUFRA
PROJECT

TIBESTI
MTNS

Darnah

Al Beida
Cyrene Plateau
JEBEL AL
AKHDAR
PROJECT
Jebel al-Akhdar

Benghazi

MEDITERRANEAN SEA

TAWURGHA
SPRINGS
PROJECT

Tawurgha
Springs

Sirte Plain

WESTERN

(TRIPOLITANIA)

SOUTHERN

Tripoli

Al-Aziziya

Jefara Plain

Jebel Nafusa

Ghadames

Wadi al-Shati

Wadi al-Ajal

MURZUQ
BASIN
PROJECT

Traghan

Murzuq

(FEZZAN)

FEZZAN
PROJECT

TUNISIA

ALGERIA

NIGER

300

km

0

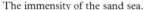
The immensity of the sand sea.

The eucalyptus tree, which has adapted well to arid conditions, dominates a plantation of olives on the coastal plain.

method of irrigation, while the numerous wells in the *saniya* (garden oases) have a high population density. The Italians in one of their colonisation schemes utilised a deeper aquifer 48 kilometres from Tripoli by means of diesel and electric pumps, which put farming in the area on to a commercial footing and allowed for further expansion southwards into virgin territory.

Recently, over-exploitation of water for irrigation has, according to Hill, caused sea-water infiltration in many of these coastal *saniya*. And ground-water has declined between 1 and 3 metres a year in areas close to Tripoli. Many pumps installed after 1960, as a result of investment, have contributed to this decline.

Eastern Libya (formerly Cyrenaica)

Here the coastal plain, where it exists at all, is generally narrower than in the Western Region, and the chief city, Benghazi, has even less rain than Tripoli—only about 250 mm a year. The thickly wooded northern uplands of Eastern Libya, which in places reach almost to the sea, are known as the Jebel al-Akhdar (Green Mountain). The extreme north of Eastern Libya provides the right environment for goats and cattle, which do well on the farms of the upland terraces. South of the Jebel al-Akhdar the land falls away and becomes mainly *hamad*, or stony desert, with a few relatively small, scattered wadis. Towards the south-east of Eastern

117

Before any form of cultivation can take place the sand dunes have to be fixed into position. Here dead grasses are used as hedges before a tree is planted in the centre of each plot. Oil and latex mulch stabilises sand, reduces evaporation, and allows the tree to root.

Alongside the Coastal road east of Tripoli in Libya, a shepherd watches his flock of sheep. The terrain reveals the increasing salinity of the area.

Good quality blood-stock obtained by importing stallions for breeding.

Unexploited springs near Beida are typical of many potential water sources.

The descent to desert can be halted. With stabilisation and sufficient water even this sand sea, which surrounds the oasis of Brak, could be made fertile.

Libya lies the true desert with its vast dunes and totally empty wastes; while in the far south, a spur of the Tibesti Mountains lies within the border of this province.

Southern Libya (formerly Fezzan)

This region consists of about 1 million square kilometres of desert, in which a number of great basins and depressions containing ground-water, often fossil water, are found. It includes the vast sand sea of Calanscio, many thousands of square kilometres of sand which is constantly shifting, with dunes that can be hundreds of metres high. Here and in the southern part of Eastern Libya are situated the great oases of the country, among which the Kufrah Oasis is the best-known and most important. In the artesian basins, as the water-tables are shallow, oasis cultivation can be carried out, dotting the desert with green islands among the stone and sand. The cultivation of dates and millet is a common factor where there are oases: the former providing subsistence for many Libyans during times of seasonal shortage of other staples.

The climate of Libya is almost uniformly arid, with a marked variety and wide alternations of temperature. In the uplands of Western and Eastern Libya there can be between 350 and 500 mm per year, the latter receiving more water, because of its promontory, than the former, which lies in a rain shadow. Elsewhere the average ranges from 200 mm to zero. Nor is the rainfall reliable; it is often erratic and there is a drought every few years, so that prediction and planning become extremely difficult. In fact, variable temperatures and rainfall, particularly in the north, are more pronounced than in Egypt. In summer, Libya is generally very hot: figures of over 50°C are not unknown in the southern desert, while even in the Jefara they can reach a maximum of over 45°C. Western Libya is more prone to higher temperatures and even greater aridity, with the fearsome and disastrous sand wind, the *ghibli*, which devastates crops and all other growth in its pathway.

AGRICULTURE

The presence of ground-water is of prime importance in determining the intensity of production in the areas of higher rainfall, and hence in dictating agricultural expansion.

Agriculture can be divided into three types depending on the physical conditions prevailing: cultivation round oases in the desert; that of Jefara and the nearby regions of the Western Jebel (Tripolitania); and the mixed cultivation pattern typical of northern Cyrenaica (Eastern Libya).

Despite the unfavourable physical features and climatic conditions, agriculture dominated the Libyan economy until oil was discovered.

Barley and olives in order to produce profitable yields require an average rainfall of 200 mm, and as these are two of the country's principal agricultural crops, in conjunction with the increasingly important citrus and tomato crops, it can be seen that water is of prime importance if development is to continue. Pastoralism is an economic necessity, but Libya had 1·3 million goats and 1·7 million sheep in 1969, and overgrazing limits the growth of the vital perennial forage grasses in areas where rainfall is between 150 and 200 mm. Esparto grass, which is used for top-quality paper production, has been increasingly phased out. In the days before oil was discovered, esparto was Libya's chief export.

As a result of the discovery of oil, agriculture is threatened by the drift to the towns, where the opportunities are obviously greater, of those workers who already form a minority of the population. The 3·8 million hectares of agricultural land, mostly consisting of meadow and pasture, have remained the same over the last ten years.

The output of an artesian well being put to practical use for the growth of beans.

In such country, harsh in the extreme, beneath the surface life-giving water may be found.

AGRICULTURAL DEVELOPMENT

Libya can be divided agriculturally into a distinct pattern governed by rainfall and potential irrigation, within the framework of which are the traditional system of agriculture on the one hand, which relies heavily on nature and is haphazardly practised, and a well-organised and planned agricultural system on the other.

The former system is prevalent in the steppe region, such as the central and southern parts of the Jebel al-Akhdar, the Cyrene Plateau, and Beida, together with some of the Jefara area and the Sirte Plain.

Libya, in common with all countries which have vast areas of desert, lives with constant and ever-increasing threat of encroachment by the sand which creeps across and envelops productive and potentially productive land. One of the best methods of 'desert control' is afforestation. Libya began to deal with this problem as far back as 1952, with a massive tree-planting programme designed to 'hold back' the Sahara. Acacia, eucalyptus, and pine trees have taken to these desert conditions. Interestingly enough, oil by-products are playing a major part in tree growth: after the saplings are planted, the sand is sprayed with an oil-based solution which settles as a protective layer over the sand, keeping it in place and giving the saplings a chance to root properly. After a year they are then sufficiently established to be able to withstand the effects of wind and erosion, and have long enough roots to bind the loose soil. Aerial photographs taken on the fringes of the Libyan deserts show how far advanced Libya has become in her bid to tame the encroachment of sand, by the criss-cross pattern of trees which predominates.

Because of Libya's poor water-resources, only a limited area of its vast territory is suitable for agricultural settlement: chiefly the coastal strips, the northern slopes of the Jefara and Jebel al-Akhdar, and the coastal and desert oases.

During the Italian occupation before the Second World War, apart from large estates specialising in tree-crops, some smaller demographic settlements were established in order to increase the number of Italian settlers, who formed communities of peasant farmers. Elsewhere in Libya, traditional habits persisted among the indigenous settled cultivators; and post-independence Governments have had to contend with social and psychological problems which no amount of money can quickly solve.

When a people are used for centuries to a highly traditional way of life, often a nomadic one, it is extremely difficult fundamentally to change their habits. This applies not only to the Bedouin, but also to those Libyans who were already settled cultivators. As a report by the United Nations Food and Agriculture Organisation on land settlement schemes said with wisdom:

> It must be borne in mind—and the National Agricultural Settlement Authority is well aware of it—that 'settler' is not a synonym for 'farmer'. To organise a land settlement scheme; to allocate a plot of land in it to each family; to give a contract to each family, to finance them in order to place them in a position where they may live and grow their first crops, and so on—all this does not mean the creation of successful settlers, nor the assurance of the success of the whole project. Continued assistance and daily extension, education and advice are absolutely essential to develop new outlooks, to show new aspects of rural work, to teach techniques, and so to produce skilled farmers out of all settlers.

To illustrate this point, when the Italian colonisation was ended, there was an immediate rush by the Libyans to occupy the farms and houses abandoned by their former Italian occupants. But several difficulties prevented the continued efficient running of these establishments. Most of the aqueducts and rain-water reservoirs had been damaged, so that water-supply for both drinking and irrigation proved to be a serious problem. Again, there was little money available in those days, and nothing to replace the Italian administrative machinery. But there were also human problems:

1. The farmhouses were incompatible with the traditions of Libyan family life, so that the houses were often abandoned in favour of shanties or tents erected beside them.

2. The Libyans were not versed in systems of crop-rotation, nor were they familiar with modern agricultural methods and equipment.

3. There were very few experienced managers and supervisors to help them maintain, let alone develop, the abandoned farms.

The problems facing recent Libyan Governments in this field can therefore be summarised as follows:

1. Rural settlement.
2. Attracting, training and supervising new farmers, many of whom had previously been nomadic or semi-nomadic.
3. Increasing the amount of cultivated land by technical development.

Irrigation of a cash crop of beans.

Orange pickers. At this experimental research station, pest and weed control, the use of fertilizers and the search for improvement in all forms of crops, poultry and livestock are part of daily routine.

Let us take these problems in turn and see what in recent years the authorities have been able to do to overcome them.

RURAL SETTLEMENT

Rural settlement and agricultural development in a country such as Libya are totally correlated. For the object of rural settlement in Libya is to create economically viable societies on arable or reclaimed lands by inducing people to come and live on these lands by gradually adapting them to a new socio-economic environment. This environment can be created through the collective effort which must be made to reclaim and prepare the land for profitable agriculture and to construct the necessary amenities for the new settlers—houses, roads, schools, hospitals, etc.

The Libyan Government accordingly allocated 17 per cent of the first Five-Year Plan (1963–68) to the development of the agricultural sector. This included an appropriation of £10 million for programmes specifically designed to promote agricultural settlement. To carry out these programmes, the Government created the National Organisation for Agricultural Settlement in 1963. The importance of demographic agricultural settlement was again stressed in the terms of reference given to the General Organisation for Agrarian Reform and the Rehabilitation of Land, which was created in 1970 with the object of 'contributing to the development of national income in the sphere of agrarian reform and the rehabilitation of land'. To achieve this purpose, the Organisation was charged with the following tasks:

A method of water retention is to divide farm land into small plots, each with an earthen bank.

Water is conveyed by gravity flow from an artesian well through this concrete irrigation channel in the Brak area.

1. To survey cultivable but unused areas and to draw up a general policy for the rehabilitation and cultivation of these lands.

2. To administer and develop all State-owned agricultural lands throughout the country, with the exception of those directly administered by the Ministry of Agriculture and Agrarian Reform.

3. To draw up the necessary programme for the execution of reclamation, cultivation and rehabilitation schemes, in order to bring about vertical and horizontal expansion of the areas under the Organisation's control within the framework of the overall development plan approved by the State.

4. To promote irrigation and reclamation schemes and projects related thereto such as digging water-wells, constructing buildings and organising such utilities, roads, power-lines and other amenities as might be required for the full implementation of each scheme. In putting these schemes into effect the Organisation might act either independently or in co-operation with other Government agencies or specialised companies.

5. To draw up suitable cycles of crop-rotation for the areas under its control, and to exploit this land economically through cultivation, irrigation, promotion of animal husbandry, and marketing of the agricultural and animal products.

6. To distribute the agricultural lands, after their reclamation, rehabilitation and division into productive farms, to farmers in accordance with the regulations and conditions laid down by the State.

7. To draw up programmes for the development of rural societies in the areas under the Organisation's control and to implement such programmes, including the provision of social and cultural amenities and other services for the guidance of the farmers who benefit from the distribution of the Organisation's land.

8. To establish and supervise agricultural co-operatives among the beneficiaries of land distribution; to encourage these co-operatives, and otherwise to take care of their affairs to the benefit of their members.

9. To draw up and carry out training programmes in all matters related to the Organisation's activities, in order to raise the skills and productive capacities of all its employees.

TRAINING OF NEW SETTLERS

According to the 1967 official statistics there were 37,988 families in the Bedouin (nomad) and 30,170 families in the semi-nomad categories. This was a labour force which could well be used to contribute towards the development of agricultural production if favourable conditions were created under which these Bedouin might settle down and form viable agricultural communities.

Accordingly a team of experts from the Food and Agricultural Organisation of the United Nations (FAO) has made preliminary studies of the Bedouin population and has formulated recommendations for the solution of their problems. They include a proposal for the settlement of the nomadic population in the southern regions in selected areas by offering them certain facilities and services and by inducing them to take part in agricultural projects.

The type of agricultural settlement which the FAO recommended for Libya was of the fragmented-holdings type, since it was more suited to the small population and the vast expanses of agricultural land, often separated by barren rocky areas which break the continuity of the farmed lands.

The 1970 Law for the Disposal of Reclaimed and State-owned Lands provided for the distribution of these lands among Libyans who were unable to earn a living wage, provided that they were farmers by profession or capable of taking up farming as their vocation. The object of this law, of course, was to create agricultural communities on the newly-reclaimed or uncultivated farmlands.

Studies conducted by FAO experts of some of the projects had revealed that the major difficulty which faced the beneficiaries, and which often undermined their willingness to settle, was their ignorance of how to manage their farms as integral production units following a regular system of crop-rotation. They also had difficulty in adapting themselves to the requirements of co-operative work and to modern agricultural methods.

The agricultural settlements were ready to receive the new farmers so far as housing facilities, water supply and technical farming requirements were concerned, but other public utilities such as schools, hospitals and other services were still lagging behind, since the construction of these utilities was the responsibility of different Government departments. It was therefore felt to be necessary that the

Poultry are cross-bred at experimental agricultural stations to obtain the bird best suited to the local environment.

settlement projects should in future be planned as a whole, including the necessary public services and amenities.

As to bringing the land up to an economically productive level, it was found difficult to determine the correlation between productivity and size of the individual farm, in view of the numerous variables involved, e.g. the variety of soils in different regions, the local rainfall pattern, and the kinds of crops grown.

On some of the old demographic settlements it was found that the beneficiaries were not wholly devoted to their farm work, but had other non-agricultural vocations which they pursued at one and the same time. As to the payment by the beneficiaries of the price of the farms allotted to them, the 1970 Law for the Disposal of Reclaimed and State-owned Lands stipulated that such payment should be made by instalments over a period of between fifteen and twenty years from the date of receiving the farm. The first instalment was to be paid out of the proceeds accruing from the first crop produced.

Training and rehabilitation courses were to be held in order to adapt the beneficiaries both vocationally and socially to the new conditions in the settlement. A balanced programme of training and instruction was drawn up, with the object of helping the nomadic and semi-nomadic settlers and their families to adjust to the new way of life, and to train them vocationally in settled agriculture and farm management. They had also to be persuaded to live in houses rather than in tents. With these ends in view, the training and rehabilitation programmes were designed to serve the following purposes:

1. To impress upon the beneficiaries the purposes and benefits of agricultural settlement, pointing out how the new way of life would raise the economic, cultural and health standards of the farmers while making them useful citizens capable of contributing to the welfare of the society in which they lived.

2. By taking new settlers to visit the model 'family farm' which was created for this purpose in the Saadia settlement scheme, to give them practical instruction in the crop-rotation system and other agricultural operations, including poultry farming and animal husbandry. Trained extension workers would then follow up with help and guidance when the farmers were running their own establishments.

3. To instruct new beneficiaries, through lectures and seminars, in the benefits of co-operative farming and in the methods of setting up and managing agricultural co-operative societies.

4. To instruct the settler families in the benefits of home economy, housecraft and hygiene.

5. To hold demonstration courses on the application of crop-rotation systems, poultry farming, horticulture, pest control and the cultivation and harvesting of field crops.

6. To instruct the settlers daily by means of films in all matters related to their new life, both vocational and social.

7. To hold literacy courses, in co-operation with the nearest school, to teach the new settlers to read and write.

8. To establish social and recreation clubs for the new farmers in collaboration with the nearest school.

Over the past seventeen years great improvements have been made in Libyan agriculture with the introduction of mechanisation, new techniques, and new methods including research projects like those at the experimental station of Sidi el-Mesri near Tripoli, which is run by the Ministry of Agriculture. Here the emphasis is on long-term agricultural development, with seeds, cuttings, and saplings obtained from many parts of the world, experimented with under Libyan growth conditions, and able to stand the type of regional climatic variations. Farmers ultimately benefit from this in better varieties of crop, which will increase their yield and boost their income. It is the type of bold plan which can help to stem the drift away from the land.

The Libyan National Agricultural Bank assists farmers with the purchase of machinery and fertilisers and equipment to replace the out-of-date methods of irrigation. For instance, the diesel pump is rapidly gaining in popularity as a substitute for the animals which have for centuries provided the power for lifting water. Other incentives offered are fifteen-year interest-free loans for those willing to develop uncultivated areas or to improve land. Support is given to co-operatives. This movement started in Libya in 1956, and has been gaining in support ever since, ranging from farms to fisheries, all of which are subsidised by the State. They pool their resources, and such items as machinery can be drawn for a small charge from State-run agricultural machinery stations. Easy terms are available for such vital equipment as well-drilling gear. Produce is also marketed in consumer co-operatives with the aim of each village owning one co-operative shop, in a bid to help solve the constant problem of inflation.

In order to lessen to some extent the dichotomy between urban and rural areas, development projects in undeveloped areas have been formulated with the object of attracting settlers to an agrarian way of life. Among these are the projects at Kufra, the Wadis Ajal, Shati, and Traghan (Hofra oasis), and Jebel al-Akhdar.

Kufra is one of the most famous oases in Libya, and has attracted scores of Arab and European travellers through the ages. It was chosen as the site of a comparatively large project for agricultural expansion because of the apparent abundance of underground water, the suitability of its climate for the growth of a variety of crops, and the availability of large tracts of level arable land.

The scheme was actually initiated in November 1968 by cultivating 100 hectares with clover (used for soil enrichment and stall-fed cattle), and a similar area with barley. By the end of 1970 the cultivated area had increased to about 525 hectares.

This successful result prompted the authorities to dig a battery of 100 artesian wells, which were completed in November 1971. Water is pumped mechanically from these wells to irrigate an additional area of 10,000 hectares.

The Kufra Project envisaged a medium-term expansion of the cultivated area to 60,000 hectares by 1976, and its main object was to increase the production of wheat, barley and livestock. According to the long-term programme, by 1980 the production of wheat and barley is expected to reach 350,000 tons, and the number of sheep to exceed 500,000.

The Wadis Ajal, Shati and Traghan Projects are utilising wadis, or dry river-beds, situated in Southern Libya and constituting the northern section of the

Murzuq basin and other lowlands that have been found suitable for cultivation.

Here, after a detailed survey of a total area of 40,000 square kilometres prior to the selection of a suitable area of 60,000 hectares for development, the initial stage of reclamation is confined to an experimental area of 3,500 hectares, for which a detailed programme has been worked out. This includes facilities for training Libyan workers in agricultural methods, the establishment of workshops for repair and maintenance, and a land research centre, the digging of wells and the construction of farmhouses on a site of 1,500 hectares with one house for every holding of 5 hectares. The cultivation of the reclaimed land follows, and selection of the farmers to whom the land will be distributed according to the conditions laid down by the State.

Some attention has already been given to the Jebel al-Akhdar area because of its agricultural importance. Recently, the potentialities of its further development have been under review, a plan for which was initiated in 1970. While additional studies were being made to determine the size of the agricultural unit that would be economically viable to provide an adequate standard of living for a farmer and his dependants, further work began that year on the afforestation of an area of 12,000 hectares as well as on the reconstruction and maintenance of 2,000 farmhouses which had been left abandoned in the region.

Other Government schemes include the building of dams across major wadis to conserve the rainwater. Such dams are seen as preliminaries to the possible creation of future agricultural settlements. Large volumes of ground-water are known, or thought, to be present in many countries in the Arab lands. Libya, with recent discoveries in her desert areas as a corollary to exploratory activity for oil, is but the latest addition to the list. Government investment has shifted perceptibly to the south of the country, where the agricultural schemes are designed to exploit the finite fossil water resources. Nevertheless, as Clawson *et al.* point out, little is known of the origin of such supplies and the existence and rate of recharge, or indeed whether the supplies could support agricultural activity of the magnitude required to form any significant addition to present production. The answers have yet to be found.

LAND TENURE

The Settlement Project lands and other lands controlled by the Land Reclamation Organisation are regarded as State property leased to the beneficiaries; but consideration is being given to the possibility of gradual transference of ownership to the farmers on a long-term hire-purchase basis. Lands recovered from the Italians and other foreigners, however, are held by the farmers on the basis of the *metayage* system, that is, the farmer pays a proportion of his revenue to the owner (in this case the Government), who furnishes stock and seed free of charge.

How far such a large but relatively empty country (the population is one-twentieth of that of neighbouring Egypt), with its oil wealth, can improve its agriculture and industry without the help of outside labour is a problem to which only time will show the way towards a solution.

Until such time as irrigation and reclamation of the land have become more widespread and developed, animal husbandry will still form the backbone of farming. Nevertheless the predominance of sheep and goats for their meat, milk and wool is on the wane, due to the increase in the breeding of cattle for dairy produce. The introduction of good breeding-stock from countries such as the Argentine is also providing a larger number of beef cattle.

The figures shown are for 1973; in the last few years, the proportion of goats and of camels has declined.

Land Use in Libya 1971 (*hectares*)

Arable land	2,377
Land under permanent crops	144
Permanent meadows and pastures	7,000
Forests and woodlands	534
Other land	165,899
Total	175,954

Principal crops
(*tons*)

	1973	1974
Barley	116,395	204,514
Wheat	41,585	67,327
Olives	94,533	149,313
Citrus fruits	27,138	20,046

Main Livestock 1973

Horses	145,000
Camels	185,000
Cattle	950,000
Sheep	8,000,000
Goats	2,300,000

Source: *The Middle East and North Africa 1975–76*, Europa Publications, 22nd edition.

REFERENCES

Terence Blunsum: *Libya, The Country and its People*, The Queen Anne Press, 1968

R. Farley, *Planning for Development in Libya*, Pall Mall, London 1971

W. B. Fisher, *The Middle East*, 6th edition, Methuen, 1971

Marion Clawson, Hans H. Landsberg, Lyle T. Alexander, *The Agricultural Potential of the Middle East. Economic and Political Problems and Prospects*, Elsevier, New York, London and Amsterdam, 1971

J. I. Clarke and W. B. Fisher, *Populations of the Middle East and North Africa. A Geographical Approach*, University of London Press, 1972

The Republic of Tunisia

Tunisia is a relatively small country (the smallest of the three states which make up the Maghreb) by the standards of those we have considered so far, wedged between the two colossi of Libya and Algeria on the North African coast and occupying some 164,000 square kilometres of territory. It is bounded on the north and east by a long Mediterranean coastline. The northern part of the country enjoys a typical Mediterranean climate which, together with its historical associations, extensive plains, and its relative closeness to Europe, makes the Tunisian Republic less harsh geographically and gives it a national identity distinct in the Maghreb.

After oil was discovered in Algeria and Libya, it was eventually located also in the far south of Tunisia (where the al-Borma and Douled fields are situated), providing Tunisia with much needed revenue for future development of the country. But compared with that of other states in the Arab World it is a small industry.

Most of the towns are situated in the coastal areas, and it is the north and east of the country that contain the majority of the population, which in 1969 was estimated at 4·8 million people. About 54 per cent of the population live in an area comprising some 16 per cent of the whole.

PHYSICAL FEATURES

Tunisia can for our purposes be divided into two regions: the populated North on the one hand and the Centre and South on the other.

The Northern Region

This area is dominated by the two easternmost chains of the Atlas Mountains, the Northern Tell and the High Tell. These are separated by the deep, fertile, alluvial plains of the Medjerda Valley, in a region which has a rainfall of 400 to 600 mm a year, and which is dominated by agriculture, particularly cereals, with an emphasis on a monoculture regime.

The important Medjerda River, which flows from west to east through low-lying plains, has two tributaries, the Beja and the Zargha, which originate in the mountains of the Kroumirie and Mogod, and the Melleque, the Tessa and the Siliana which drain the Tell mountains. It is the only perennially-flowing river in Tunisia and is, as we shall see, of the greatest economic importance. A delta has formed at its mouth—the only one in north-west Africa.

In northern Tunisia, the climate is of the Mediterranean type, with a warm damp winter followed by a hot dry summer. It contains the wettest region in North

Africa, the Kroumirie highlands, where the average annual rainfall is as much as 1,500 mm, though in most of the region the average varies from 375 to about 1,000 mm. The High Tell area is characterised by pine forests and by a number of high, level plateaux, while cork oak flourishes in the Northern Tell.

The most important Tunisian cities are situated in the northern area. Tunis, the capital, lying near the historic Phoenician capital of Carthage, has been influenced by its eminent predecessors and is the centre of political, cultural, commercial and industrial activity, with a first-class port. It contains four-fifths of the country's industry. The northern area also includes a number of other commercial centres, such as Mateur, Nabeul, Menzel Bourguiba, and Menzel Temime.

The Central and Southern Region

This can be subdivided as follows:

The Eastern Coastal Plain

This is a sizeable area known locally as the *Sahel*, nearly 320 kilometres long and varying between 30 and 70 kilometres in width, extending the entire length of the eastern Tunisian coast from Boufisha in the north to Mahares in the south. The presence of the sea moderates the temperatures in this region, and rain, though scarce, is sufficient for the cultivation of Mediterranean crops, and for extensive raising of stock as well as afforestation.

The Steppes

West of the Sahel and south of the High Tell lie the low and high steppes, which are characterised by only slight rainfall, while the occasional water-courses that cross them flow only after heavy rain and are usually dry.

Annual rainfall is under 350 mm for the region, with even less in the south at Gafsa (about 150 mm). The aridity of this steppe area is accentuated by the hot, dry winds which blow from the Sahara in the summer unimpeded by any land obstacle.

Soil is of the steppe type, and vegetation is sparse. Some Aleppo pine forest growth is supported. A livelihood is possible from sheep, goat and camel rearing, with some cereal cultivation in the lower steppe region.

The Southern Depression and the Southern Desert

South of the high steppes, the land gives way to a broad depression containing great seasonal salt lakes, or *shotts*, salt marshes, and palm grove oases, the most important of which is the vast Shott Djerid, which lies more than 15 metres below sea-level and is normally covered by a salt crust. Under the salt crust water is found, and date-palm groves are irrigated from artesian wells in this *shott* region. Sand dune encroachment is prevented by screens. Date-palm leaves act as very good windbreaks, to protect crops and nurseries.

This area gradually gives way to nearly 52,000 square kilometres of full desert in the far south, where the climate is arid in the extreme. This is the Tunisian Sahara, practically uninhabited except for pockets of oases where dates and olives are grown.

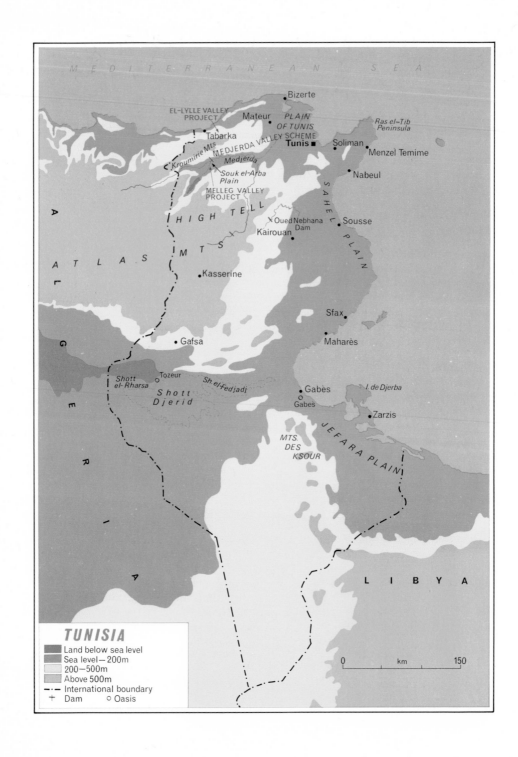

MEDITERRANEAN SEA

Bizerte

EL-LYLLE VALLEY
PROJECT
Mateur
PLAIN
OF TUNIS
Ras el–Tib
Peninsula

Tabarka
Kroumirie Mts
MEDJERDA VALLEY SCHEME
Tunis ■
Soliman
Menzel Temime

Medjerda
Souk el–Arba
Plain
MELLEG VALLEY
PROJECT
Nabeul

HIGH   TELL
Oued Nebhana
Dam
Sousse

SAHEL PLAIN

Kairouan

MTS

Kasserine

A T L A S

Sfax

A
L
G
Gafsa
Maharès

Shott
el–Rharsa
Tozeur
Sh el–Fedjadj
Gabès
I. de Djerba

Shott
Djerid
Gabès
Zarzis

E

MTS.
DES
KSOUR
JEFARA PLAIN

R

LIBYA

I

A

TUNISIA

Land below sea level
Sea level—200m
200—500m
Above 500m
-·-·- International boundary
+ Dam   ○ Oasis

0   km   150

Maintenance of an open earth channel, off-take of a Roman well, by manual labour in the Tozeur area. This type of channel requires continual attention.

Olive grove in the region of the Sahel. The two-year flowering cycle of the olive makes it economic to inter-plant with other cash crops such as broad beans, grape vines and apricot trees. There are some 44 million olive trees in the country. A tree takes from seven to ten years to mature, and its life is about 250 years. Harvesting usually takes about three months.

Experiments in Sfax have proved that a better yield is normally obtained when olive trees are well spaced out than when they are planted close together.

AGRICULTURE

An estimated 70 per cent of the Tunisian population depends on agriculture, which accounts for about 40 per cent of the national income. In general, in years when rainfall is heavy, agricultural production realises a surplus for export, but during the years of drought it decreases to such an extent that half the livestock have been known to die of thirst and starvation. As an illustration, the average annual production of wheat between 1948 and 1957 was some 670,000 metric tons; this decreased to about 165,000 tons in 1961, and in such disastrous years as this the importing of grain becomes necessary, which naturally has a grave effect upon the balance of payments.

The area of arable land in Tunisia is about 9,000,000 hectares: some 72 per cent of the non-desert area. The potential sown pasturage and fodder crops under rain-fed conditions, according to the FAO, are: 1,000 hectares of permanent pasture and some 35,000 hectares of fodder or pasture; with some 500 hectares under irrigation conditions. This improvement of pasturage is another progressive step with which the Ministry of Agriculture is concerned.

The northern part of Tunisia is the most fertile and productive area in the country. The abundance of natural water resources aids the cultivation of grain and fruit in the plains and valleys, and of olives on the high ground surrounding Bizerta. Oil is refined here, too. Aleppo pine and cork oak trees grow in the mountain areas. Sheep and goats graze in the rich pasture-land. The planned economic development of these areas includes afforestation, protection against destruction of the saplings by erosion and indiscriminate grazing, expanding the cultivation of fruit trees, increasing drainage projects in the plains, and developing the cultivation of summer crops such as tobacco, maize, millet and animal fodder.

The most productive areas of the whole country lie in the Medjerda Valley, in the plains surrounding the city of Tunis and in the Ras el-Tib Peninsula. It is noteworthy that most of these areas were previously in the hands of the European settlers because the best Mediterranean crops are cultivated in them: wheat and barley are grown everywhere, olives feature in Mernak and Soliman, vineyards in Mernak, Boufisha and Rafraf, oranges and vegetables in Nabeul and Menzel Bourguiba, and flowers and tobacco in the Peninsula.

Some collective farms have been established in the Medjerda Valley, and farmers are guided by the Government over the selection of the crops to be grown and the deployment of manpower. The American Technical Aid Project has contributed about 60 per cent of the finance needed for this scheme.

The Medjerda Valley and the Plain of Tunis account for most of the country's production of wheat, its most important single crop, and of its vines and citrus fruit. Barley is more resistant to drought than wheat, and is grown in many parts of Tunisia (mainly the central and southern regions). It will thrive in areas where there is little rain and still produce a good yield for both man and beast. Research at Tunisia's seed co-operative society investigates seed quality of crops and those best suited to climatic conditions, disease resistance (including drought and cold) and yield; this work includes barley and wheat. Great efforts have also been exerted to increase the cultivation of peaches, pears, plums and nuts, which are more

The fight against the encroachment of sand. Here dunes are being planted with deep-rooted grasses to prevent further shifting.

profitable than grain crops and less unpredictable in terms of annual production.

The rainfall in the eastern coastal plain is sufficient for growing olives on its sandy soils, as well as for the cultivation of grain, vines, figs, almonds and pomegranates. In fact, the most characteristic aspect of this area is the extensive cultivation of the olive, which can be seen in districts round Sfax, Sousse, Zarzis and Djerba. The groves almost resemble forests, with the exception of those around Sousse and the island of Djerba, where a family type of agriculture is practised in small well-managed plots. Olive oil is a major export, Tunisian oil being famous for its quality.

Under the system of co-cultivation the Tunisian farmer is given the right to own half the area of land on which he grows olives, in return for his attending to the groves for a period of fifteen years, that is, until they begin to yield. Development plans include research into the unknown reason why there is a tendency towards lower yields from olive trees; into conservation of soils, prevention of erosion (shelter-belts lessen the effects of erosion by wind), and exploitation of underground water, when not too saline, for agricultural purposes. The deep rooting of trees makes full use of soil and underground water resources. In the south, the

The inter-planting of citrus with olive trees is common practice in the north-east, where the valleys are fertile and the soil is particularly suitable for citrus cultivation (*left*).

Barley is generally grown by the dry-farming method but in the Matmata region *meska* can be seen which utilize the run-off water from the hills (*right*).

Marc, the residue of compressed grapes.

distance between olive trees is about 24 metres in all directions, while it is only 9 metres in the north: yet the yield per hectare in the south is higher than that in the north, often because the trees in many of the northern groves are very old and are not given the necessary care.

In addition to the expansion of olive production, there is an increase in the cultivation of vegetables, fodder, pomegranate and palm-trees, using the limited irrigation water in the coastal region.

As already seen, dry farming is possible in the lower steppe area, though large areas support only the wild growth of esparto grass, both for export and for conversion into paper at a newly-established mill in Kairouan. An increasing number of Tunisians live in this area, however, which is highly suitable for livestock grazing: cattle-breeding is a major industry here, accounting for about 20 per cent of the total agricultural income of the country.

Farther south, the extremes of temperature and the absence of vegetation make it impossible for anything but a sparse nomadic population to survive, except in places where underground water makes cultivation practicable. A notable example of this is the date-producing oasis of Tozeur. Other date-palm oases are those of Dakkash, Nokta, el-Hamma and Gabes, and the Government has schemes for developing date-plantations in the south.

A camel is used in this traditional method of drawing water from an ancient well. The principle is the same as that of the *shadouf*; the water raised is transferred into a channel for irrigation.

141

THE IMPROVEMENT OF AGRICULTURAL PRODUCTION

Climatic difficulties are the obstacle which lies in the path of progressive farming practices. Nevertheless Tunisia has adapted her farming to the climate and to what she has in the way of resources. Sfax is an example of practical application in the extension of olive cultivation. Dry-farming practices are paying off, as a result of experiments carried out in research stations and of furtherance of education in agriculture; and both Tunisian and French farmers, backed by the Ministry of Agriculture, have taken up the challenge of experimenting with new techniques and crops (melon, almond, and peach, for instance), often forming co-operatives to take advantage of mechanisation schemes. The value of all this can be seen in the increase in production of cereals, vegetables, vineyards, and tobacco.

Date orchards in the oases are often interplanted with other crops such as figs, vegetables, and olives. This tendency to plant date trees close together may be responsible for a reduction in yield.

The measures that Tunisia has taken to increase agricultural production and exports, to eliminate unemployment in the countryside, to face the continuous increase in population, and to raise the people's standard of living can be summed up as follows:

Modern Irrigation Methods

Introducing modern irrigation methods in Tunisia has helped to expand the areas of cultivated land and to protect some of these against the flood dangers resulting from the fluctuations in rainfall. It has also helped in the vertical expansion of agricultural production through raising productivity. Irrigation has been practised in Tunisia for centuries. The Roman system of maintaining full absorption of rain-waters, coupled with conservation of these to their fullest extent, survives to this day in the regions of Sfax and Zarzis. Maximum irrigation benefits can be obtained if the run-off from the nearest watershed is concentrated on a given area only and not dispersed widely. (Keeping land free of weeds means less loss of moisture through transpiration.) This is particularly true of dry-farming practices. Sometimes run-off water is led from the hills to the olive groves by trenches known as *meska*; these can be seen for instance in the surrounds of the Matmata hills.

Unfortunately, however, the success of irrigation in Tunisia has so far been limited by the basic scarcity of water. The main source of water for Tunisia's irrigation needs is undoubtedly obtained from surface run-off water which happens only during a short season annually, with great fluctuations from year to year. Thus in the construction of storage reservoirs sufficient leeway has to be allowed for heavy rainfall which would result in disastrous floods were they filled to capacity at such a time. The northern part of the country is the only region in which rain falls in quantities sufficient for growing grain and for permanent pasture, fluctuating between 400 and 1,200 mm. The rate of rainfall in the central part of the country, with the exception of its coastal area where there are many plateaux, is very small, not exceeding 200 mm. The FAO has suggested that in central Tunisia pumping of underground water could greatly increase irrigation potential and thus help the nomadic population towards a more settled way of life. In the south, the rainfall is very scarce or non-existent. The whole problem is greatly exacerbated, as we have seen, by the variation of annual rainfall especially in the centre and the south.

Furthermore, there are no long and perennially-flowing rivers in Tunisia except the Medjerda, which has been harnessed for maximum benefit. The Medjerda Valley Scheme is a development of major importance in irrigation and flood control, by means of three dams which are regulated by the Medjerda Valley Administration.

In the low valley some 40,000 hectares have been irrigated and drained, with an additional 200,000 hectares brought into cultivation by soil conservation methods. The main tributary of the Medjerda, the Melleque, has a large dam which controls the flow of the waters. The high valleys are also being exploited, as fertile soil is of vital importance in a country where there is not a great deal of water and one of whose problems is a sharp increase in population.

The aim is to increase agriculture and to diversify its production with either irrigated crops or dry farming (about 500,000 hectares). The former includes stepping up the production from market gardens, while the latter proposes to move on from a single crop regime to growing such crops as sugar beet, flax, cotton, and forage crops.

The problems are the same as those posed by irrigation methods in other countries, namely inadequate drainage (the Medjerda Valley Administration has established a network of canals) where the land is low, salinity of irrigation water, and evaporation.

There are other methods applied in Tunisia to increase the cultivated area; these include the construction of reservoirs and dams at the valley-junctions in the

A network of *flumes*. The weight of the water is held by a bridge to decrease the deadweight effect of the gate and also to decrease the water pressure. Windbreaks round the reservoir are intended to stop silting.

Water regulators.

A vertical-flap regulator-gate and stand-by reservoir, showing the outlet with the main channel open.

levelled areas, especially in the Tunisian Zahr area, and the exploitation of subterranean water on a wide scale in the depression zone and the Jefara Plain. There are, too, thousands of wells scattered throughout the land which serve as natural storage points for rain-water to be used for small-scale irrigation, and the Irrigation Administration, which is affiliated to the Ministry of Agriculture, is exerting considerable efforts in this direction. It provides financial aid to people who work in water-prospecting operations; and through this aid as well as through the efforts of individuals, new wells are being drilled.

Commercial Crop Expansion

The cultivation of wheat is generally limited to the areas where the annual rainfall is around 400 mm. In areas where there is a higher rainfall, citrus fruits and vines are grown. The Government's Four-Year Plan, 1965–68, was geared towards industrial expansion. Further development schemes for agriculture have depended to a great extent on foreign aid, which has been forthcoming as can be seen from Tunisia's achievements not only in irrigation but also in crop diversification.

The Government have also given considerable thought to the serious problem of soil conservation, and a policy of afforestation is in train. Contour ploughing and the planting of trees on about 2 per cent of each farm are further measures which the Government has adopted against erosion. In the sphere of animal husbandry, research is going on into methods of improving the breeding of cattle and sheep; while schemes are in progress to persuade more of the sheep and cattle herdsmen to become settled farmers.

144

LAND TENURE

Land tenure in Tunisia has a complicated but interesting history, and no picture, however brief, of the agriculture of the country is complete without some account of land-tenure systems and their evolution, especially as in Tunisia, for the first time in this book, we touch upon the effects of French colonisation.

Land tenure in Tunisia was traditionally divided into the following: private land owned by individuals, *habous* land endowed to Islamic foundations and trusts (*wakfs*), *oroush* land jointly owned by various tribes, and finally land owned by the State.

When the French occupied Tunisia their main concern was to encourage the immigration and settlement of Europeans in Tunisia, without regard to the legal difference between a protectorate and a colony. As it was difficult for the protectorate authorities to apply an official colonialist policy at a time when it wanted to prove to French public opinion that the occupation of Tunisia did not entail vast expenditure by the French Government, their efforts were restricted during the early stages to encouraging free colonisation. But they came up against the problem of finding legal loopholes to ensure the rights of colonists who wanted to exploit the land with complete freedom.

In 1885, it was decided to apply a system of real estate ownership similar to that employed in Australia, where there were no indigenous civilised inhabitants already settled on the land with formal systems of tenure. According to the French law of that time the new owner of a piece of land could formalise his title to it by registering it, after delineating its borders and proclaiming his ownership at a special court established for that purpose. A copy of the title-deed was kept in a register at the courthouse as a record in case the owner wanted to dispose of his land by selling or hiring it out. Under this system, the new owner could rest assured that no other person could claim the ownership of his land.

But in spite of these facilities, it was noted that few French people were encouraged to emigrate to Tunisia. Those who did so were mostly big capitalists who bought extensive areas of land and leased it to the original inhabitants or to other Europeans of various nationalities. Italians, in fact, formed the majority of the workers, and many of them became farmers in the northern and eastern section of the country, although they were encouraged to obtain French nationality because the Italian Government was opposed to the inducements of the French. The Maltese also formed a community in commerce and farming.

The policy of expansion of European immigration was very dangerous, particularly as Tunisia was under-populated at the time of the protectorate. The number of Tunisians at that time was less than 1·5 million, most of whom were concentrated in the coastal areas, while the other parts of the country were very sparsely populated indeed; the population increase did not proceed at the same rate as in Algeria, for instance. Even by 1946 the population had reached only 3 million; and, as we shall see later, the seizure by the Europeans of a considerable part of the agricultural land, most of the industries and all the mineral wealth, hampered the economic progress of the Tunisians. European rule concentrated on the better endowed northern and maritime regions to the detriment of the arid regions of both the centre and the south of the country. Urbanisation was expanded, as were the ports of Tunis, Sfax, Bizerta, and Sousse. In fact, four-fifths of the Europeans lived in the towns.

Using under-population as a pretext and in an effort to balance the flood of Italians into Tunisia, the French authorities gradually turned the protectorate into a *de facto* colony ('colonie de peuplement'). The first step taken towards this end was the issue of a decree in February 1892 providing for the annexation by the State of the fallow land—a step which resulted in the sequestration of vast areas around Sfax, where the Europeans could thus infiltrate into the olive groves which were the main source of livelihood of the original inhabitants.

The second step was the seizure of the land held by the *habous*. In order to mitigate the effects of the shock thus administered to the considerable number of Tunisians who lived on these lands, the authorities temporarily avoided the sequestration of the civil trusts. But, as neither form of endowment is open to sale, the French sought a loophole in Islamic jurisprudence to justify the seizure of these lands by the settlers. They found such a loophole in a weak passage in Malek's doctrine, which they used to enable the settlers to seize 1,000 hectares per settler per year from the vast areas of the trust-lands in return for a nominal annual rent. But the settlers, with their enormous influence over the protectorate authorities, did not respect even the minor provisions restricting their leases and they ceased, from 1905, to pay any rent. In 1904 they proposed to sequestrate also the land of the civil trusts; but it was realised that the beneficiaries of these lands, who were mostly Tunisian notables, would, if they were deprived of their wealth, become staunch enemies of France and that this would greatly strengthen the Tunisian nationalist movement.

The third step was the expansion of nationalisation. At the time of the imposition of the protectorate in 1881 state ownership did not exceed 100,000 hectares. But within a very short time the European colonists occupied this land and started looking for other areas. Various decrees were issued bringing pastures and water-springs into state ownership, thus creating disputes between the authorities and the tribes which benefited from the pasturelands in northern Tunisia. In order to eliminate any claim by the tribes to the possession of the pastures, a decision was issued by the court to the effect that a tribe does not have a corporate legal identity whereby it can claim ownership. The protectorate authorities pursued this policy of sequestration until the amount of land held by the Europeans had reached after the Second World War about 800,000 hectares. Here it should be noted that agricultural colonisation in Tunisia was characterised by a sort of centralised feudalism; four capitalist organisations owned about 140,000 hectares, that is, 23 per cent of the total.

Since Tunisia became independent, considerable changes have taken place in the system of land ownership, and a complete transformation in land distribution has been effected. The Government nationalised the land of the civil trusts, amounting to more than 400,000 hectares, divided it into small plots and distributed it among the farmers. It also took over the *oroush* land, which amounted to about two and a quarter million hectares and was publicly owned by the tribes, again divided it into small plots and distributed it among the Bedouin in an attempt to encourage them to settle and work as sedentary farmers.

In 1957 the Europeans owned approximately 740,000 hectares, but gradually this area decreased to 455,000 hectares in 1959 and to 400,000 hectares in 1961, with France helping Tunisia to buy this land. In March 1963 an agreement was concluded between the two countries whereby the Tunisian Government bought a further 150,000 hectares from the Europeans during that year. By 1964 the majority of the colonists' farms had been expropriated and nationalised, and distributed among Tunisian farmers. Thus the departure of the colonists was on a gradually diminishing scale and unhurried.

Nevertheless, the Tunisian economy is still closely knit to that of France (small oilfields do not make sufficient impact), but is undergoing at present a period which might be termed quietly transitional with no economic revolution.

REFERENCES

M. Y. Nuttonson, *The Physical Environment and Agriculture of Morocco, Algeria, and Tunisia, with Special Reference to their Regions Containing Areas Climatically and Latitudinally Analagous to Israel*, American Institute of Crop Ecology, Washington D.C. 1961

J. I. Clarke and W. B. Fisher, *Populations of the Middle East and North Africa : A Geographical Approach*, University of London Press, 1972

Sir L. Dudley Stamp and W. T. W. Morgan, *Africa : A Study in Tropical Development*, 3rd edition, John Wiley and Sons, New York 1972

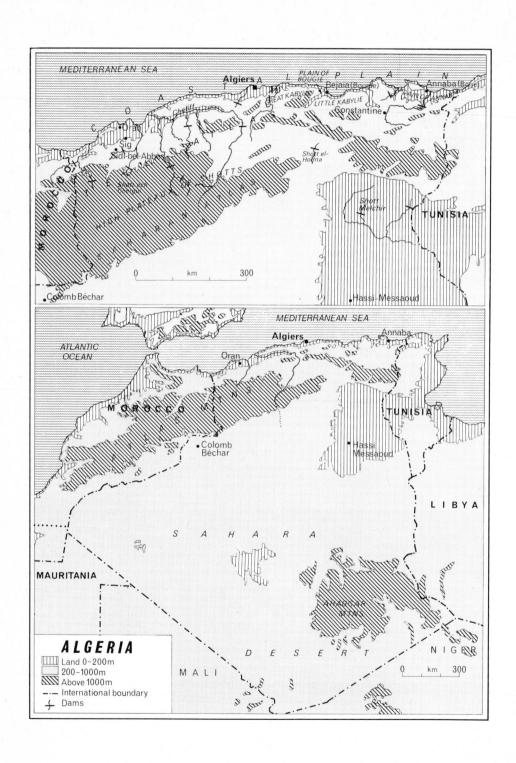

MEDITERRANEAN SEA

Algiers

PLAIN OF BOUGIE

Bejaia (Bougie)

Annaba (Bône)

GREAT KABYLIE

LITTLE KABYLIE

Constantine

Chelif

Sig

Short el-Houna

Sidi-bel-Abbès

SHOTTS

Shott Melchir

TUNISIA

0   km   300

Colomb Béchar

Hassi-Messaoud

MEDITERRANEAN SEA

ATLANTIC OCEAN

Algiers

Annaba

Oran

MOROCCO

TUNISIA

Colomb Béchar

Hassi Messaoud

LIBYA

MAURITANIA

S A H A R A

D E S E R T

NIGER

MALI

ALGERIA

Land 0–200m
200–1000m
Above 1000m
International boundary
Dams

0   km   300

The Democratic Republic of Algeria

Algeria occupies about 2,466,800 square kilometres. This huge area is divided into two main regions: Northern Algeria, the smaller but 'settled' area, with 295,000 square kilometres, 12 per cent of the whole, with about 94 per cent of the population, while Southern or Desert Algeria makes up the rest. There is a line of oases on the Saharan side of the Atlas ranges which forms a convenient demarcation between the two.

Northern Algeria is a narrow belt along the shores of the Mediterranean with a coastline of almost 1,000 kilometres; and it is in these lowlands that most of the population of just over 12,000,000 people live. Only about 700,000 inhabit Southern Algeria. The country is bordered on the east by Tunisia and Libya, on the south by Mali and Niger, and on the west by Morocco and Mauritania.

Population movement and the results of the 1966 census would indicate in recent years a natural increase rate of 3·5 per cent, which could give Algeria one of the highest population growth rates in the world. This, in the wake of the tremendous upheavals, violence, and disorders which preceded her independence in 1962 (Algeria was a French possession), could have seriously affected planned economic development.

Nevertheless by 1980 the Algerians believe they will have transformed themselves from a country of agriculture (confined mainly to a strip on the edge of the desert) into an industrial economy, very competitive, and able to utilise natural resources of oil and liquid gas which are vitally needed. Many other useful mineral deposits are found, but these tend to be overshadowed by the world demand for both oil and gas. However, the need for agrarian reform has not been neglected, as we shall later see. But the latter type of reform is not easy and success does not lie round the corner.

Northern Algeria consists of narrow coastal plains and valleys, broadening here and there in the areas where the main towns, Oran, Algiers and Annaba are situated, and is backed almost immediately by the Atlas Mountain range. The mountain system in Algeria is made up of three wide bands, or zones. To the north is the Tell Atlas, a series of massifs and plateaux at heights in places of over 2,000 metres and separated from one another by deep valleys and gorges which divide this area into distinctive units. South of the Tell Atlas there is a great level plateau, known as the High Plateaux of Shotts. A *shott* (in French *chott*) is a shallow lake, often saline and frequently not permanent. In spite of their proximity to the ocean, the Plateaux of Shotts (600–900 metres high) are isolated from maritime influences, with consequent extremes of temperature and likewise progressively differing vegetation and agriculture. South again of the High Plateaux lies the Saharan Atlas which, in the north-east corner of Algeria, north of the Shott el-Hodna, merges with the Tell Atlas and ceases to be a distinct range.

Southern Algeria stretches for almost 1,600 kilometres into the depths of the Sahara and most of it consists of considerable stretches of *erg* as opposed to true desert, with exceedingly sparse vegetation, if any, and with only occasional oases where underground water makes life possible for this vast territory's few inhabitants, amounting to only five persons per square kilometre.

The climate of the coastal plain and the Tell Atlas region of the north is Mediterranean: hot and dry in the summer, warm and wet in winter. The rainfall is highly variable, from an annual average of more than 800 mm down to less than 200 mm, depending on altitude and position. The uneven rainfall together with high rates of evaporation lead to the curious situation that not only are Algeria's rivers, flowing from the Tell Atlas to the Mediterranean, not perennial, but they are short in length and their water is frequently either low or non-existent; while an area can be subject to severe flooding by particularly heavy rainfall occurring in a very short time.

In the High Plateaux and the Saharan Atlas a more arid type of climate prevails, with an average rainfall of 400–600 mm, and a stunted, steppe type of vegetation. Below the mountains, the rainfall steadily decreases from less than 200 mm to almost nothing in the south, while maximum shade temperatures of 54°C are on record.

The result of a short-lived torrent in the desert, showing exposed rock face interspaced with fertile soil brought down from the mountains. Utilization of such deposits could increase the agricultural potential of the land.

LAND TENURE

The land problem featured very largely in the prolonged and bitter conflict between the French colonists and the Algerian farmers before independence. France had wanted to create in Algeria a 'new France' for the settlement of French farmers. This was scarcely easy, since most of Algeria is desert, only a very small percentage of the land is arable, and such of it as was cultivated was in Algerian hands. Reconciliation of French objectives with the actual position then obtaining in Algeria was impossible, and one of the two parties had to give way. The Algerians fought hard for their land but eventually had to withdraw to the interior where the soil is less fertile and cultivable land less accessible.

Thus nearly all the modern farms of Algeria were until 1962 owned by Europeans. A small number of individuals and companies owned large plantations. Two thousand eight hundred landowners owned 200 hectares each, and more than one-third of the farms were over 10 hectares in extent. European farms were concentrated in the country's most fertile areas—the coastal plains and the northern sector of the High Plateaux where water sources are abundant.

Modern methods of agriculture were not adopted by all the European farmers, as many of these were small landowners who had emigrated from the poorer regions of France. Much of the colonists' development of the land was, however, with its good communications network, ably carried out. Consequently, the average production per hectare in the areas owned by Europeans was higher than in those owned by Algerians, and this was due, in the first place, to the prevalence of favourable natural conditions, such as good soil and water supply, and in the second, to greater education, knowledge and skill.

A quarter of the annual budget is now devoted to education, leading to a full programme of schooling by 1980. Algeria is closing the gap left by the French in the disciplines vitally needed not only for agricultural expansion but for industrial programming as well, which has been filled to some extent by the use of foreign expertise and aid.

The Government actively promotes agriculture and livestock husbandry in the form of experimental farms, schools of agriculture and horticulture, and co-operatives to foster the increase of the agricultural potential.

Four-fifths of the French settlers had left Algeria by the time independence was declared. As these formed the backbone of the intensive farming community with a market geared to the French market, with exports particularly of wine (viticulture forming the developed modern branch of the sector within the general economic pattern), this was a very disruptive factor in agriculture which Algeria had to face. Sacrifice in living standards, in order to reap the benefit from oil revenues for the development of Algeria, is a cause which most Algerians are prepared to accept. And industrial and agricultural development is going ahead with an austerity in personal terms which might be emulated with advantage by the western world.

AGRICULTURE

Some eight million people live on the land, which is mostly harsh and inhospitable and difficult to cultivate. Contrary to Saudi Arabia, where an emphasis has been placed on agricultural development, Algeria has opted for a more industrialised economy. Nevertheless there is a 'revolution' going on on the land, in spite of peasant opposition to change and modern methods (primitive methods accelerated the erosion process in the poorest soil areas). Agriculture, which engages more than half of the population, thus accounts for only 10 per cent of the country's gross national product. Phosphate production supplies the needs of the country for fertilisers.

In general terms, concentration of the people continues towards the coastal, or Tell, cities and plains, while a more rural belt of 'high density' can be found eastwards through the Grande and Petite Kabylie to the Collo Massif. This, because it is an area of traditional Berber settlement, was left to its own devices by the French as a distinctive region of high population within the coastal Tell mountains.

Principal Crops Under Cultivation (Hectares)

	1972	1973
Wheat	2,403,000	1,100,000 (estimated)
Barley	720,000	450,000 (estimated)
Vines	no figures available	1,200,000 (estimated)

The Coastal Plain

This plain comprises a number of pockets or valleys lying between the Atlas Mountains and the sea. These fertile valleys, found at intervals along the coast, are dominated by the Atlas ranges with their typical corrugated profile, elevated plateaux, and well-watered Mediterranean slopes. Where alluvial soil is present and water obtainable, it is a natural consequence that settlement should be more extensive, as productive use is made of the fertile plains for market gardening (to supply the ever-increasing home market), grapes, citrus groves, cotton and tobacco. Irrigation has been practised since Roman times, with the French increasing the cultivated area by a system of water control that was strictly supervised. Thus although these plains around Oran (the central port and city), taking in the Cheliff Valley, Algiers, Bougie and Annaba, amount to about 3 per cent of the total land, according to the 1966 population census they still account for more than half the population density. The Tell region coastal strip, some 150 kilometres wide with a rainfall averaging between 400 and 800 mm, has a natural Mediterranean vegetation.

Numerous wadis cross the coastal region, with the majority of the country's streams flowing only when there is heavy rainfall. As the rainfall decreases inland, the local streams dry up during the summer, but can be torrential, erosive too, during the rainy season, hence the characteristically steep river banks in Algeria.

All rivers share the same formulae: surface water (run-off) discontinues; water 'disappears', flowing underground; dry wadis are supplied by springs; evaporation takes its toll, as do the alluvial deposits which gather momentum downstream and cause problems at barrages. Some of the rivers and streams are used for irrigation, but cannot be relied on to provide sustained water when required. Barrages are a better solution because the water can be stored behind them during the time of the rains and diverted for agricultural needs throughout the Tell. Smaller domestic supplies of water are obtained from shallow, deep, or artesian wells.

With independence also came the headache of the continuance of the viticulture industry without the necessary experts to run it, coupled with a decline in exports to France—France having enough wine for home consumption. It takes years to establish a vineyard, and high labour output. Although grapes need less irrigation water than, for instance, oranges, future policy may move towards cereals as a substitute. The flooding of the export market with citrus fruit from other countries may well influence Algerian policy to consider what is best for home consumption in a country with a phenomenal rise in population.

The production of citrus fruits has increased considerably in the past twenty years, and future prospects for the citrus industry are brighter than those for the vineyards because of the increase in exports to Western Europe.

Vegetable-growing is concentrated in the areas near Algiers, Oran, and Annaba, where there is an outlet market in the developing industrial cities. The early production of vegetables in Algeria gives the country a marked economic advantage in that its produce appears in French markets well before the French-grown vegetables. The favourable growing conditions can provide a two-crop annual cycle: one dependent upon irrigation, the other on the winter rains. The increase in the urban population has led to a comparable increase in the production of citrus fruit, vegetables and olives, and to the equipment of Oran, a vital international port, with refrigeration plant for such produce.

The main socialist sector of farming, which was formerly the French settlement areas, lies between Algiers and Oran. Reclamation of much of the marshlands has been carried out with irrigation schemes in operation on about half a million hectares and barrages acting as storage reservoirs for river water (many have fallen into disrepair and the government is now surveying water potential as part of its agrarian reform policy). Algeria has fewer opportunities for the furtherance of irrigation because of lack of water.

In common with other North African countries she has a comprehensive locust-control programme, to destroy locusts in their breeding grounds before they can devastate vast areas.

Sugar beet is grown in grape-farming regions, as it does not need irrigation. The question of crop rotation is of great importance in this type of cultivation. A typical pattern of rotation could be as follows:

> Wheat
> Sugar beet *or* fallow
> Sugar beet
> Wheat

A sugar beet experimental station is situated at Sidi-bel-Abbes.

The need for a policy of crop rotation, control of grazing and resettlement of population in areas where there is not even a subsistence level of cultivation possible are aspects already being considered in planning agrarian development. Changing a way of life is not easy for those who have lived in one area for generations. The terracing of mountainsides to prepare them for cultivation is a costly process out of reach of many farmers. Continuous stands of forest are rarely found in Algeria. One of the main functions of these is to act as protectors from erosion and water losses, in addition to supplying timber and cork. Forests in Algeria are managed by a Forestry Service, which arranges protection from over-grazing with consequent loss of trees, and organises afforestation schemes.

The coastal plain, as we have seen, is the site of the main cities of Algeria. The foreign settlers, of whom 80 per cent were town-dwellers, have mostly left since independence. In 1960, only 22 per cent of the native population lived in the towns,

Camel terrain.

154

and one fifth of this number were unemployed. The city of Algiers, whose population was 834,000 in 1960, is the chief city commercially, industrially, culturally and politically. Nevertheless, it does not account for so much of the country's trade as one might imagine, owing to the distances between the producing areas on the long coast and to the competition of the other ports. Oran, with a population of 393,000, has a better natural harbour than that of Algiers. In addition to the richness of its surrounding land with its agricultural products, it exports iron ore and is linked to Colomb Bechar by railways. The latter city lies in the centre of a rich mining area in the south. Annaba, moreover, also exports minerals, while Bougie is the most important oil port in Algeria.

The coastal towns are linked by a modern well-equipped railway line but the means of communication from south to north are less advanced than they need to be.

The Tell Atlas

In Arabic the word 'tell' means hill. The Tell Atlas hills and mountain ranges contain the comparatively well-watered coastal region, and valleys, lowland basins and plains which support considerable forests and fertile agricultural land. The Tell Atlas rises from Oran in the west to 2,500 metres in Great Kabylie in the east.

The eastern Tell is drier, so the population tends to concentrate in the rainier hill areas, the traditional country of the Berbers, whose livelihood depends on sheep, wheat, figs and olives.

Eastern Algeria has mainly cork oak forests, with conifers and grazing land. Elsewhere many of the forests have been destroyed throughout the centuries by man and beast, and erosion plays havoc on the bare mountain-sides, which, despite rainfall, cannot amass sufficient soil for even the *maquis* type of vegetation to gain a footing.

Where there is adequate water and alluvial soil, cereal crops, fruits and vegetables flourish, with the grape intensively cultivated, and the date-palm in the oases.

The merino wool-producing breed of sheep predominates with the goat on the *maquis* and semi-arid regions, where they are able to flourish on the natural forage available.

The High Plateaux of the Shotts

In the interior the High Plateaux take over southwards from the Tell Atlas, stretching into Western Algeria. These plateaux, which are in the main made up of semi-arid steppes, form valuable grazing for sheep and cultivation of esparto grass. They are also known as the Plateaux of the Shotts (averaging 800 metres in height), with a width of approximately 200 kilometres in Oran in the west, stretching southwards to the Saharan Atlas. The interior contains huge landlocked saline *shotts* (basins) which form salt lakes during the winter rains, draining away in the summer to form salt crusts.

Soft wheat and black oats, superior in yield and resistant to disease, grow on the drier and less fertile Constantine high plains, often by dry farming methods, while the Plateaux of Shotts form grazing country with some esparto grass for paper-making on those dry lands which continue across the Saharan Atlas to the margins

Harsh and inhospitable landscape: the Ahaggar massif.

of the desert. Irrigation could greatly extend agricultural potential in this region.

The Saharan Atlas

This lies south of the High Plateaux, and tends to be generally higher and less broken than the Tell Atlas, flanking the Saharan Desert and stretching from Morocco to Tunisia. It is made up of a series of mountain chains which ultimately break down into unconnected ridges.

Southern Algeria

To the south of the Saharan Atlas stretches the immense desert area, the surface of which is made up of bare rock, pebbles, and other debris, only about 10 per cent consisting of *erg* or sand dunes. A typical feature of this region is the huge mountain massif of the Ahaggar. In the desert, nomadic pastoralism with the raising of goats and camels is the major way of life (except for the oases where water is available for the cultivation of irrigated crops). It is estimated that two-thirds of Algerians living in the desert are farmers settled in the oases and the remaining third are nomadic shepherds. The Algerian oases contain four million date-palms, which produce an average of 67,000 tons of dates yearly, 18,000 tons of which are of a quality suitable for export. Horticulture is also important and various kinds of fruit and vegetables are cultivated in the oases. However, the whole of this production is consumed locally.

Some attention has been given to the possibilities of reclamation of land in the desert in order to help the population problem. The existing oases obtain their water either from underground sources, or from deep wadis after heavy rains. Sometimes small dams are constructed across the streams in order to encourage the infiltration of more water under the soil. However, only those crops which need little water, such as barley and wheat, can survive in these conditions.

Most oases depend on wells which are dug deep to reach the subterranean waters. Some oases, fortunately, have access to artesian wells. However, the water must be drawn up either by human effort or by using animals. Mechanical water-pumps are rarely used. Curiously enough, the high price of petrol is the main reason for the infrequent use of pumps; and this in an area floating on oil, so to speak!

A kind of horizontal tunnel or *foggara* is used for drawing water from the underground water-preserving layers to plots which are of a lower elevation. Some of the *foggara* are very well constructed, but maintenance difficulties have led to the deterioration of some oases using this method of obtaining water. In some areas, where the level of subterranean water is near to the surface of the earth, the soil is dug deeply enough to allow the roots of the date-palms to receive their water directly from the underground source.

One dominant problem in these oases is that of protecting cultivated plots against the encroaching sand, a process that demands continuous labour and care. Revenue from agricultural production rarely justifies the huge efforts that are exerted in providing water and preserving some of the oases. The assessment of the possibilities of expansion of irrigation in the desert cannot be carried out unless and until a thorough study of the subterranean water-resources has been completed. Examples are many of the oases in which agriculture expanded after wells were dug but deteriorated again because of the eventual draining-away of available water-resources. In other areas, digging wells for expansion in agriculture in certain oases caused a decrease in the water supply of other prosperous oases a few miles away; for both were using the same subterranean source.

Since 1930, a number of deep wells have been dug in the Algerian desert, but these were vastly expensive, since to prepare one well to a depth of 1,000 metres costs over $280,000 (£100,000). Sometimes the water has been found so far below the desert that its use would be totally uneconomic; while in other places the water was too saline to be used.

However, from the research now available it should be possible to double the cultivated area in the Algerian desert. Yet, even if all the available water-resources were used, the total irrigated area would only scratch the surface of the great desert and would be too small to cause any basic change in the economic status of the desert and its inhabitants.

THE SOCIAL ORGANISATION OF AGRICULTURE IN ALGERIA

In order to draw a clear picture of the agricultural conditions in Algeria, we must realise the importance of agriculture to its economy, despite the trend towards industrialisation (see page 149), and understand the current organisation of both the socialist and the traditional sectors, and the potentialities available in each. Thus it will be possible to evaluate how the co-operative movement operates in this field.

Apart from land owned by the State, land-ownership in Algeria is divided into two distinct sectors: the socialist sector and the traditional sector.

The Socialist Sector

The socialist sector is composed of the land which was owned by French landowners, of whom there were 22,037, holding about 2,920,000 hectares. These holdings were and still are the most fertile and highly-productive lands in Algeria. A high proportion of agricultural income comes from the products of this sector.

In July 1962 the French landowners evacuated the land and left it in the hands of the agricultural workers who used to cultivate it. Work continued on these farms as before, and farmers organised themselves in the beginning by electing some of their number to administer and carry on the work in the same manner as the French had done.

The area of land taken over by Algerian farmers in July 1962 was about 880,000 hectares, cultivated as follows:

Crop	Hectares
Grain	700,000
Vineyards	120,000
Fruit	23,000
Vegetables	18,000
Cotton	12,000
Olive-groves	5,000
Tobacco	3,000
Rice	500
Total	881,500

Since this was the only practicable and immediate way of preserving the agricultural wealth of the country, the Algerian Government approved the actions of the agricultural workers in the abandoned French holdings. But in March 1963 laws were promulgated to formalise the procedure of self-administration. Thus the system known as self-administration came into being in Algeria. In fact, it was the display of initiative by the workers in continuing production that saved the economy of Algeria from a positive catastrophe. In March 1963, the Algerian

Government nationalised an area of nearly 800,000 hectares previously owned by big French landowners, and on 7 October 1963 it nationalised the remaining area of one million hectares which had also been in the hands of the French.

The Traditional Sector

This sector still relies on more primitive methods of cultivation, using implements such as wooden ploughs with iron ploughshares, drawn by horses, mules and oxen. Where ploughs cannot be used, as in the Kabylie region because of the steep slopes, hoes are used to break the soil surface. The seed is then sown by the broadcast method. Improved seed varieties and fertilisers are slowly being introduced into this sector. Harvesting is done with the aid of animals and without modern machinery, but again time is changing the habits of generations of farmers who are slowly finding easier methods of crop cultivation within their grasp as an indirect result of oil and gas revenues.

Of the eight million people or so who live within the scope of the traditional sector, about two million are not farmers, working often in urban occupations; and of the other six million only two million work in agriculture, the rest being herdsmen.

Land Use in Algeria

	(*Hectares*) 1970
Arable	6,240,000
Permanent Crops	552,000
Meadows and Pastures	37,416
Forest	2,424
Irrigated land	191,542

Source: FAO *Production Yearbook.*

THE PROCESS OF SELF-ADMINISTRATION

This system was used in Algeria to achieve efficient administration and to avoid bureaucracy, on the one hand, and on the other to create an atmosphere in which workers and farmers would feel that the revolution was theirs, that the public interest was identical with their own interest, and that the good of the country was their own good, so that they would exert every effort to increase production and to safeguard public property which was placed in their hands.

This system is also regarded as a training ground for workers and farmers in which they are taught political and social consciousness, and how to take responsibility. The self-administration workers manage collective organisations both in their own interests and for the benefit of the country at large.

The experiment of self-administration in Algeria is distinguished by the fact that it started from the bottom. It was the workers who took the initiative in managing farms and factories left by the French, and who elected committees from among their number for that purpose. Workers endured great sacrifices in the early months for the sake of preserving and safeguarding these organisations and of ensuring their continued operation. They worked without any pay and exerted tremendous efforts to repair broken machines. Self-administration committees were also given the necessary funds with which to pay the workers' wages and to market their products.

It was also necessary to establish specialised organisations which would encourage the application of the results of this experiment in other fields. In the matter of supplying the collective farms with the necessary money, tools, and seeds and fertilisers, the State depended on the steel organisations and particularly on the Kabir Organisation, which constitutes the main administrative and technical organ in the National Organisation for Agrarian Reform.

An organisation for general planning and guidance, affiliated to the Council of Ministers, was formed at first under the title of The National Office for the Protection and Administration of Vacant Properties, but in March 1963 its name became The National Office for the Development of the Socialist Sector.

Besides these organs, other national organisations played an important and decisive role in supporting and ensuring the success of this experiment. The General Confederation of Algerian Trade Unions took a major part in stirring up the enthusiasm of the workers and in explaining the decisions of the self-administration authorities to all those concerned.

The main tasks of workers in the self-administration organisations were to preserve the assets placed in their hands, and to increase production. It was their right freely to sell and purchase the tools and machines put at their disposal; it was also their duty to husband the resources of the organisation they were operating and administering.

Before it distributes incomes, the workers' board contributes to the expenses fund of the organisation, determines the sum necessary for developing this organisation and deposits this amount in its special investment fund. The board of workers also determines the share necessary for the workers' social fund. This is set aside for building houses, running health and education services, and organising entertainment for the workers in their leisure time. The workers themselves also contribute to the various co-operative and municipal funds.

The organisations' properties are not transferable, and if any worker quits an organisation he has no right to any of its funds. There is a sharp distinction between the personal income of the worker, that is to say, his monthly pay and his share from the general annual income, and the organisation's assets and reserves. Apart from the workers' collective responsibility to preserve this wealth and to increase production, it is also incumbent upon them as a body to ensure conditions of full employment, while the State appoints a technical manager who is responsible for defining the maximum number of workers needed on each farm. One of the main

The threat of encroachment by the desert is ever present.

Verdant irrigated valley showing a *flume* system.

Sweetwater spring with a typical *erg* formation in the background.

conditions of working in self-administration organisations is that the worker should have no other source of income but that which he earns from his work. In order to contribute to the process of full employment, workers in all farms subscribe a fixed sum to the 'fund for the balance of employment'.

The system of self-administration faced some difficulties in the technical field owing to the void left by the withdrawal of European technicians, in addition to other commercial and financial problems. In order to tackle this major problem, the National Organisation for Agrarian Reform laid down a large-scale programme for rapid vocational training. The fact that most of these difficulties were overcome during the first year of self-administration helped enormously to bring into being a new approach to agriculture in a newly-developed Arab country.

The main areas of farming in the socialist sector lie in the western Tell between Algiers and Oran, where modern farming methods flourish. Although some of the country round Oran and Algiers is very fertile, it still has to be developed. Much of the coastal area was reclaimed, while irrigation was needed on the plateaux and plains in other areas, which resulted in the development of small irrigation schemes. The lack of significant rivers seriously handicaps Algeria's agriculture, where reliance has to be placed on rainfall and underground water resources.

AGRARIAN REFORM

The unequal distribution of land was the main reason for the low standard of the rural masses, and for their inability to improve their methods of agriculture and to contribute to the economic development of Algeria.

The area of arable land in Algeria is, as we have seen, relatively small owing to the presence of mountains, plains and desert. The land now under self-administration constitutes about one third of the arable land. The fact that it was previously owned by colonists indicates that it is the best-situated and most highly productive land. Therefore, it is cultivated under the supervision of important agricultural organisations, equipped with modern means of production which help it to produce most of the agricultural exports and to be the main source of supply to the home market.

Land-ownership in the private sector is subject to different laws from those of the socialist sector. Consequently, we find that there is still an unequal distribution of land among farmers. The latest studies show that there are many big landowners, and that much arable land is owned by urban people or others who do not practise the profession of agriculture.

In fact an unequal land distribution already existed when France first occupied Algeria in 1830. But the French occupation helped to widen the gap in the ownership and distribution of the agricultural areas. The following figures show the areas of land seized by the colonists:

Period	Area (hectares)
1840–1860	365,000
1860–1880	315,000
1880–1900	243,000
1900–1920	1,327,000

Meanwhile, the enactment of financial laws, direct and indirect pressures, and the effects of economic crises were the main factors which forced Algerian families to sell their land to the colonists. An area of about $2\frac{1}{2}$ million hectares of the best land was taken from the Algerian farmers to become French property. Furthermore, there were some areas of land which were forcibly taken from the farmers who had hitherto either owned or cultivated it. Protective laws fell far short, too, with regard to forests and pastureland, while the effects of local wars impoverished some of the defeated tribes.

As a result of the farmers' migration from the rich and fertile areas which they had previously cultivated to mountainous and barren areas in the south, they were obliged to cultivate sterile soil with insufficient means, and to try to reclaim forests and pastures without the tools or money to do the job properly.

These facts lie behind the current density of population in the relatively infertile regions, the gradual decline of the productive capacity of these regions, and the inefficiency of farmers when, after independence, they started to use modern methods of agriculture.

The French authorities tried to rely on certain tribal chiefs to act as agents between their administration and the population. Those chiefs were paid for their services, either by giving them land or by offering them a legal document justifying their ownership of a part of the land collectively owned by the tribes. Others, benefiting from their privileged position in the regime, succeeded in making immense profits from trading with the occupying power, and were thus able to buy land from their fellow citizens who had lost everything.

Thus the methods of agrarian reform in Algeria are different in form and content from those in the other Arab countries we have considered. Future prosperity for Algeria does not necessarily depend on her finite oil and gas deposits, but on the utilisation of the revenue from these products as stepping-stones towards agrarian progress and development in the face of a tremendous increase in population.

REFERENCES

Annales Algériennes de Géographie, 5ème, No. 9, 1970, L'Institut de Géographie de l'Université d'Alger

Sir L. Dudley Stamp and W. T. W. Morgan, *Africa: A Study in Tropical Development*, 3rd edition, John Wiley and Sons, New York 1972

M. Y. Nuttonson, *The Physical Environment and Agriculture of Morocco, Algeria, and Tunisia, with Special Reference to their Regions Containing Areas Climatically and Latitudinally Analogous to Israel*, American Institute of Crop Ecology, Washington D.C. 1961

André Tiano, *Le Maghreb entre les mythes*, Presses Universitaires de France, 1967

The Times, Special report: Algeria, Parts one and two, November 1974

Spring flowers briefly in the foothills of the Grande Kabylie Mountains in Algeria.

Contrasts in Algeria: Saharan Atlas landscape.

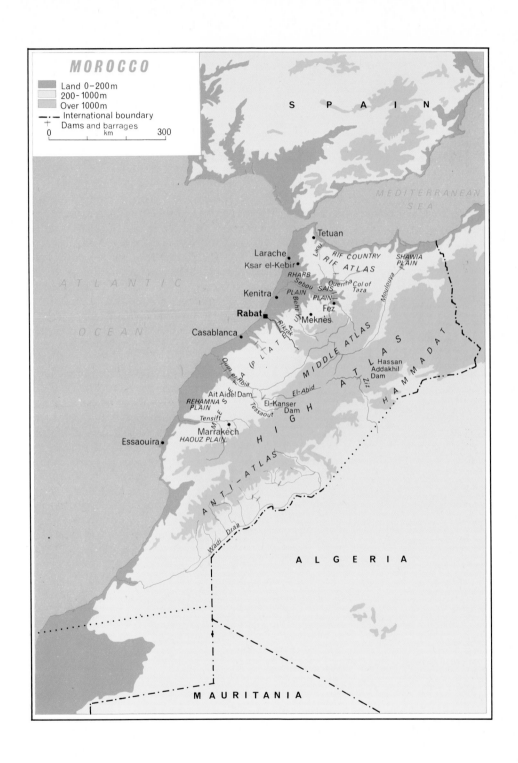

MOROCCO

Land 0–200m
200–1000m
Over 1000m
— ·· — International boundary
Dams and barrages
0   km   300

S P A I N

MEDITERRANEAN
SEA

A T L A N T I C

OCEAN

Tetuan
Larache
Ksar el-Kebir
Laou
RIF COUNTRY
RIF ATLAS
SHAWIA
PLAIN
RHARB
Sebou
Querha
Col of
Taza
SAIS
PLAIN
Kenitra
Beht
Rabat
Rhdou
Fez
Meknes
Moulouya
Casablanca
Oum el-Rbia
(PLAT
MIDDLE ATLAS
Hassan
Addakhil
Dam
Ziz
HAMMADAT
Ait Aidel Dam
El-Abid
El-Kanser
Dam
REHAMNA
PLAIN
Tessaout
HIGH
ATLAS
Tensift
Marrakech
HAOUZ PLAIN
Essaouira
ANTI - ATLAS
Wadi Draa

ALGERIA

MAURITANIA

The Kingdom of Morocco

The Kingdom of Morocco, about 458,730 square kilometres in extent (slightly smaller than California), lies at the northwest corner of the African continent and is the westernmost of the three countries—Morocco, Algeria and Tunisia—which are collectively known as the Maghreb. This term comes from the Arabic phrase 'Gezira el-Maghreb' or 'Island of the West'. Although it is very much a part of the Arab World, Morocco has been less influenced than other North African countries even by the Arabs, much less by Romans in earlier days and by Frenchmen in modern times. This is only partly due to its distance from the Arab heartlands: it is also caused by the massive mountains of Morocco's interior, which until recently have maintained the isolation both of the country as a whole and of the different ethnological communities within it, comprising Arabs, Berbers, Rif tribes, and such like. The highest and most rugged ranges of the Atlas system are what dominate Morocco. They form a barrier between the fertile plains and the north and west plateaux, on the one hand, and the semi-arid regions of the Sahara which lie to the south and east, on the other. The country possesses both an Atlantic and a Mediterranean coastline of considerable extent; and the population of Morocco in 1971 was estimated at over 15,000,000, making it now third in order of population in the Arab World after Egypt and the Sudan. The high rate of population increase is typical of many of the countries of the Middle East, but it is particularly interesting to note the exodus of a large number of Europeans and Jews during the last twenty years—Algeria is the only North African country to have had a larger mass emigration of Europeans. The 'mushrooming' of large and small towns during the same period is also of some importance.

Morocco became an independent nation in 1956 after a long period as a Protectorate of France and Spain.

PHYSICAL FEATURES

In Morocco the Atlas Mountain system is at its most elevated and rugged. The system here can be divided into four separate mountain chains or massifs interspersed with low-lying plains and upland plateaux. The Rif Atlas of the north forms a coastal border; these mountains, forming natural barriers, are hard to penetrate and are chiefly inhabited by Berber farmers. Immediately to the south of the Rif lies the Middle Atlas, which is flanked by the basins of the country's two chief rivers, the Moulouya flowing into the Mediterranean and the Oum er-Rbia flowing into the Atlantic. Most of the main rivers flow west into the Atlantic, and many have dams to facilitate irrigation schemes (see map).

The magnificence of the High Atlas Mountains. Mountain tracks suitable only for pack animals make the trading of commodities between the Atlantic and Saharan Morocco possible.

South of the Middle Atlas, the High Atlas forms the most elevated range which rises in places to over 4,000 metres and is covered with snow in winter. The northern slopes, rising above the Atlantic plain in the north, are covered with forest and *maquis*, while the slopes overlooking the Sahara Desert to the south consist simply of bare rock. The most southerly of the Atlas Mountains is the Anti-Atlas range, which is lower and less forbidding in general than the other massifs.

The Rif, Middle and High Atlas enclose the only really extensive lowland part of Morocco, which contains the Rharb Plain and the valley of the River Sebou to the north, and the plains of the Meseta, Rehamna, Djebilet, Tadla and Haouz in the south.

The vast region covering the entire pre-Saharan area in the south, which is arid, and the dry eastern Moulouya River valley is sparsely populated, containing only 6 per cent of the population; yet it covers about one-third of Morocco. There are a number of oases in the pre-Saharan region which are well watered from water-courses originating in the Atlas range. Here are the largest Moroccan villages, sited along the Draa, Ziz, Todra, and other valleys, each with its typical *ksout* fortress dominating the skyline. Here, too, cultivation of dates, vegetables, and cereals is to be expected where there is a good supply of water.

As most of the ground-water, fertile soils and rainfall are found in the western area of Morocco, it is a natural consequence that the majority of the population can also be found there. The oases which lie south and west of the mountains support only a very small proportion of the remainder as, naturally, life is harder in these situations. In this way Morocco is following a pattern very much in common with other Arab countries; a densely populated urban zone coupled with vast areas of sparsely peopled and inaccessible country.

This rock formation of the lower Anti-Atlas range is typical of the area, where barren slopes and gorges predominate, with cultivated settlements and palm groves extending into the desert wherever water is present.

The map shows how Morocco has been able to make use of her rivers to increase the extent of her irrigation.

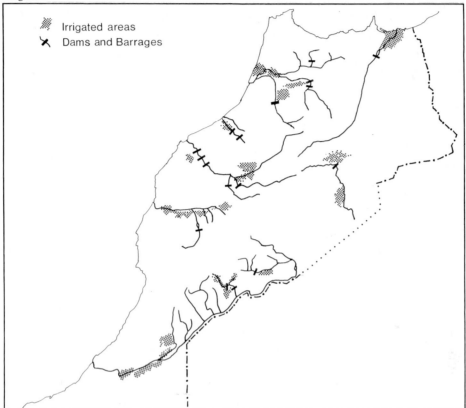

Irrigated areas
Dams and Barrages

There is a typical Mediterranean climate in north and central Morocco, which becomes more arid as one moves southwards and ends in almost desert conditions. In the far north, mean annual rainfall is over 760 mm with a short summer drought; most of the Middle Atlas, the High Atlas and the northern half of the Atlantic plain have between 380 and 760 mm of rain, while the southern part of the Atlantic plain and the Anti-Atlas enjoy only 200 to 400 mm per year. The High and Middle Atlas mountains can be described as acting as a collecting 'tank', with their western slopes catching rainfall from the south-westerly and westerly air masses which accumulate from the Atlantic.

Nevertheless, rainfall is irregular, and the variations in climate have a great impact on agriculture. At present between 5 and 10 per cent of farmland is under irrigation. (See map above.) Compared with other North African countries Morocco has a good water supply, including the large rivers already mentioned. There are also many smaller rivers, all with development potential for irrigation and hydro-electric power production.

An open irrigation channel showing clearly the type of topography suitable for the site of a dam. An orange grove where irrigation water is available contrasts dramatically with the scrub growth in the foreground.

A concrete dam constructed at the neck between vertical hills. An upstream lake serves as a reservoir.

These northern slopes of the High Atlas Mountains contrast strikingly with the bare southern slopes. The upper slopes are forested with evergreen oak, and cultivation is possible lower down. Erosion of soil is prevented by terracing, and olives and vines can be grown.

AGRICULTURE

Agriculture is the key to Morocco's economy. It presents a fairly complex picture which it may be easiest to look at regionally.

The Sebou Valley

This is one of the richest parts of the country, where population densities vary considerably. The lower basin of the river is known as the Western Plain; this is an old northern gulf filled up with alluvial deposit from the River Sebou. Swamps once covered a large part of the alluvial plain, but European farmers settled in a great part of this land until it became one of the most important areas of wheat and barley cultivation in the country. Citrus fruit is widely grown here, and accounts for 40 per cent of the country's total production. The area cultivated in citrus fruit in the Western Plain has considerably increased in recent years: from 11,000 hectares before the Second World War to more than 65,000 hectares today. Great efforts have been made to reclaim the areas of swamp that remained; 40,000 hectares north of Rabat have been irrigated and turned into one of the most productive agricultural regions in Morocco. Unexpected floods, however, present a threat to agriculture in the Western Plain. In 1962 and 1963 widespread floods caused vast damage and ruined most of the crops.

The middle basin of the Sebou River, known as the Saïs Plain, was once a lake which became a flat and fertile plain when the river water drained away. Modern farming is practised here on a large scale for production of cereals and vines. It, too, is one of the most densely-populated regions in Morocco. Many European plantations are found around Fez and Meknès, the most important cities of this area. The ancient Islamic city of Fez is the religious capital of the country, while Meknès is an important trading and communications centre. To the east lies the Col of Taza, a narrow passage in an area where the Middle Atlas is only 3 kilometres from the Rif.

The Atlantic Coastal Plain

This is a narrow strip which extends from Rabat to Essaouira (Mogador), ranging in width from 30 to 95 kilometres. It is divided into two parts: the sandy Atlantic coastal plain, and the inland area of black soil that is in the main eminently suitable for the cultivation of cereals and vegetables. The usefulness of the plain for cultivation is limited to some extent by the dry climate to the south, and there are places that must depend on irrigation. This is obtainable, fortunately, as rivers

There are many varieties of date. Here 'golden' dates are still attached to the palm branch.

Drying the date harvest in the sun.

The Arabs of Tetuan, who inhabit the north-west end of the Rif, use traditional farming methods.

flow into the plain, notably the Regreg, the Oum er-Rbia and the Tensift. Above Rabat and Kenitra, there are wide areas of dense forests from which the country obtains its cork—Rabat possesses one of the biggest cork factories in the world. The cork-oak forests have been widely destroyed by excessive cutting, grazing and burning in the past, but recent Government efforts at reafforestation are meeting with success, although it is inevitably a slow process. In some areas, the cork-tree has additional significance since its roots and foliage help in the fight against soil erosion.

On the Atlantic coast, with many ports, there is a big fishing industry which allows Morocco to enjoy second place, after Portugal, in the world production of sardines. Morocco also produces other kinds of fish, notably anchovies and tuna, and has many fish processing centres.

The country's major ports are situated along the coastal plain, though few of these are natural harbours, partly because Morocco's rivers bring down to their mouths a great deal of sediment. The construction of the artificial harbour of Casablanca by the French has greatly improved the situation. Casablanca now handles about 75 per cent of the country's sea trade and, with a population of over a million, has become one of the largest cities of Africa. The French also created Rabat as the capital city with a large services sector.

The bark of mature cork oak trees has been stripped by hand, a process which occurs at intervals after the trees are between eight and eleven years old. The cork layer is then three to five centimetres thick. At present there are some 310,000 hectares of these valuable trees.

The Meseta

Behind the coastal plain and south of Marrakech lies the Meseta Plateau, with an average height of 300 metres. This is a dry region, practically devoid of trees except at the northern edge where cork forests are found. Most of the Meseta territory is inhabited by semi-nomads. Cattle and sheep-rearing is the basis of the region's economy, as it is of most of the mountain regions, but the fluctuations of the rainfall can make cattle-breeding extremely hazardous. The breeder may lose 40 per cent of his stock in dry years, through natural causes or forced slaughter owing to lack of water and of pasture. For this reason sheep and goats are the most common forms of livestock as they are more able to seek out available water, while cattle, horses and mules are reared as pack animals.

Agriculture in the Meseta is limited to the areas where rainfall is reasonably plentiful, chiefly in the valleys where river-water can be used for irrigation, and around Marrakech where subterranean water is to be found. To the south the plain of Haouz is known for its scanty rainfall and hot, dry summer winds, but snow from the Atlas peaks supplies it with a certain amount of water, both surface and underground. This water is put to use through a network of subterranean channels which are very costly both to construct and to maintain. The area also contains forests of palm-trees while olives, vegetables and citrus fruit are grown in some places. The main crops of the area are, however, wheat and barley.

The Rif Country

The Rif area is about 250 kilometres long and 100 kilometres wide. It has never received much agricultural attention, since the Rif tribes with their own dialects live according to their ancient traditions, lack of resources constantly threatening them with famine. Local variations of soil and rainfall lend an importance to temporary migration.

Water is plentiful in the western part of the area overlooking the Atlantic Ocean, particularly in the valleys, which are fertile and support large irrigation schemes. Forests used to cover most of this region but they have been destroyed except in a limited area to the south of Larache, and agriculture and pasturelands have been substituted for trees. Larache lost a great deal of its economic importance following Casablanca's development and prosperity. One of the most significant cities of the region is Ksar el-Kebir, an important grain market.

The northern Rif consists of a range of mountains varying in height between 1,800 and 2,100 metres. Its northern border is dissected by swift-running rivers. Most of the population is employed in agriculture, cultivating grain, fruit, vines and tobacco, and their villages are situated at the foot of the mountains. The most important town is Tetuan which lies amidst olive groves only 10 kilometres from the sea.

The eastern Rif is elevated and the population-centres lie mainly on the coast, becoming less dense as one moves southwards. To the north, the people work as settled farmers, growing fruit and tending their cattle. To the south, they are mostly nomadic tribes of Berber origin, whose wealth depends mainly on their sheep, although they have some horses and camels. They practise very little agriculture.

The lower basin of the Moulouya River, which lies in the extreme east, is the richest in potential of the Rif areas.

Semi-nomadic Berber tribesmen after wintering in the valleys shift to higher slopes in the summer to pasture their flocks often on scrub growth of low, drought-resistant bushes and thin pasturage.

A sisal plantation and pack camel. Sisal, a tough fibre used for making twine, is obtained from the leaves of the *Agave sisalana* plant.

Fine blood-stock on a farm devoted mainly to horse-rearing.

The Eastern and Southern Plateaux

To the south and east of the Atlas Mountains stretches the semi-desert area of Morocco, mostly composed of the Hammadat, which are arid rocky plateaux suitable only for cattle-breeding and pasturelands. The valleys at the foot of the mountains contain palm-forests, while the oases in the area are greatly overcrowded. The north-eastern edge of the region borders on Algerian territory and is composed of high plateaux with a semi-arid climate, where large numbers of sheep and camel are bred and a modest crop of esparto is gathered.

176

LAND OWNERSHIP

At the time that Morocco became independent, the agricultural system was twofold: characterised by a small, up-to-date sector dominated by expatriates such as the French, and a large sector (about 80 per cent) of the rural population, technologically much more backward and relying on basic traditional methods with a much lower level of production. The modern sector has since passed into Moroccan control.

The French colonialists did not encourage the immigration of agricultural settlers, owing to their confrontation with the indigenous upholders of Moroccan political rights and long traditions. But the imperialist wave was stronger than any resistance, and 1,000 agricultural settlers entered Morocco, in addition to a very large number of artisans, businessmen and tradesmen in the cities. These immigrants under colonial rule owned nearly a million hectares of the 4,450,000 hectares under cultivation in 1950.

The French developed a network of modern transport facilities to connect areas and principal towns—so vital to agricultural development. The modern port of Casablanca can be attributed to their ingenuity. The problem with Morocco, as in other countries, was the need for modernisation and economic development, which in turn required the services of trained people, who happened at the time to be French. Morocco, like other protectorates, found its administration, therefore, overrun by Frenchmen and given a distinctly French character.

The door was eventually thrown wide open to immigrants and settlers. Settlers from Algeria began to exploit the eastern part of Morocco, in addition to those who came from France and settled in the Shawia Plain, the most fertile area in the Moulouya Valley, where they took over large areas. European plantations began to spread around Fez and Meknès but the settlers did not dare go farther inland until the area was made secure.

Another obstacle facing agricultural imperialism in Morocco was that most lands were owned by the religious trusts (*wakfs*) or held in common by the tribes. The French did not wish to arouse public opinion by taking over the trusts, so they directed their attention to the tribal lands. In April 1919, they issued a decree permitting the exploitation of non-cultivated tribal lands in return for a nominal rent, but this system did not attract many settlers—it was probably too dangerous at the time.

The settlers' capacity for production greatly exceeded that of the indigenous landowners. They had every financial, technical and political advantage, because agriculture in Morocco depends on a balanced and accepted system for the distribution of water. As a result of the administration's bias in favour of the French owners when the question of water-distribution arose, many disturbances took place. Medium-sized holdings were the norm among settlers, though there were a few large estates.

One of the major characteristics of the land tenure structure has been its inequality of ownership—8 per cent of the owners controlled 60 per cent of agricultural properties. Also 50 per cent of the farmers owned less than 1 hectare of land each. There is still much maldistribution among farmers. There are many very small farms owned by Moroccans, while land on lease is governed by an exceedingly complicated system. Sixty per cent of the rural population own no land at all and work as labourers or on a share-cropping system whereby the farmer receives only one-fifth of the produce. It is estimated that, taking all farm-workers together, there are 200 days in each year when they do not work at all, while 50 per cent of the rural manpower is employed for only four months in the year.

Various agrarian reform programmes have, however, been put into operation which include the collective treatment of land owned by the government with state control over the development of these properties; collecting individual farms into co-operatives; finishing irrigation projects already begun but not previously completed; soil conservation by use of fertilisers and the prevention of erosion by replanting trees and grasses and such like; and mechanisation and promotion of the livestock industry.

Morocco has an abundance of phosphate rock, phosphates being among the prime ingredients of fertilisers. As the world demand for food increases, so does the need for greater amounts of fertiliser, which Morocco can supply at her price.

In spite of the noticeable increase in the total production of citrus fruits, vegetables and wine in the past few years, Moroccan agricultural production has still not kept pace with the population increase. The individual's share of the wheat and barley harvests is not much greater than it was thirty years ago. Between 1952 and 1969, the Moroccan population increased from 8·6 to 15 million; a rate which means that the population will double every twenty-three years. Emphasis since independence has been placed on the development of agriculture and reafforestation.

Under the United Nations Development Programme, technical assistance in agricultural development was given to Morocco, and included a study for the planned economic development of the Sebou basin; pilot agricultural training in selected rural zones of the Western Rif; forestry education and training; and development of dry farming areas. Operation Engrais was a significant government effort to stimulate agricultural development in the rain-fed areas, with a view to wheat improvement. France has given technical assistance to agriculture and reafforestation projects, and has participated in the reorganisation of the Ministry of Agriculture, giving assistance to the new Agronomic Institute in Rabat by such methods as staffing and curriculum outline.

The semi-arid nature of large areas of Morocco, in addition to the tremendous variability of rainfall—droughts often followed by disastrous floods—resulted in the construction of several major dams to even out this problem (see map).

The aim is also the replacement of low-return crops (for instance, cereals) with much higher-return crops such as citrus, cotton, tomatoes and sugar beet in the newly irrigated areas.

Insufficient data on the five major irrigation projects mean that it is possible only to estimate the area of land at present irrigated. Probably 155,000 hectares are covered in this way, and about 250,000 irrigated by older systems of canals. Canals and general land development have tended to lag behind dam construction. Nevertheless, agro-industrial complexes in the project areas are going into production: sugar refineries, citrus packing and storage plants, and such like.

It is estimated that a further 930,000 hectares could be made cultivable if irrigation were available. Nevertheless, studies showed that dry land agricultural areas were easier to improve and were better suited to the production of cereals, using new high-yield varieties and making increased use of fertilisers. If this could be done, obviously the level of Moroccan agricultural production could be enormously increased. Irrigation projects are vital to Morocco because the country is unable to provide adequate food for its existing population in years of drought, while the population increases by 300,000 annually. There is also a great need for the efficient and regular maintenance of canals and barrages. Such projects and aims also run up against problems of raising the necessary capital, and plans for their implementation have to be trimmed to the extent of the cash available.

With the rising trend in population and the harsh environment in the rural areas, where often harvests are irregular and 32·9 per cent of the families working in agriculture own no land of their own, the bright lights of the cities attract the country dwellers away from the land.

A pre-cast concrete flume for conveying irrigation water on land where there is no gradient, as in the Plain of Sidi Smaine. This system of canal irrigation requires no maintenance; there is no water percolation, and no silting. The pine trees act as windbreaks and help to prevent damage to the flume when it is empty.

The failure of some of these projects obviously does not prove that they cannot succeed in the future. There have undoubtedly been notable achievements in this field. The Ait Aidel Dam on the River Tessaout, completed in 1970, is expected to increase the irrigated area of the Haouz Plain from 3,000 to 30,000 hectares, while in 1971 the Hassan Addakhil Dam on the Ziz River was completed. These are no mean achievements. But the key to any development is education, and only very little real or swift success can be achieved in an illiterate agricultural society strongly adhering to its old customs and traditions.

Morocco's main objectives are not only to improve existing facilities but to develop and allot some 400,000 hectares of new agricultural land; to make better use of extensive dry-farming methods, hand in hand with a step-up in training and research programmes; and to irrigate one million hectares by the year 2000. Irrigation programmes are vital to the production of crops for export. Prior to the boom in phosphates, agricultural produce accounted for more than 50 per cent of Morocco's earnings from exports.

REFERENCES

Antonio Gayoso, *Agricultural Development Strategies in Morocco, 1957-70*, Agricultural Economist Agency for International Development
J. I. Clarke and W. B. Fisher, *Populations of the Middle East and North Africa : A Geographical Approach*, University of London Press, 1972
Richard M. Brace, *Morocco, Algeria, Tunisia*, Prentice Hall
N. Barbour, *Morocco*, Thames and Hudson, London 1964

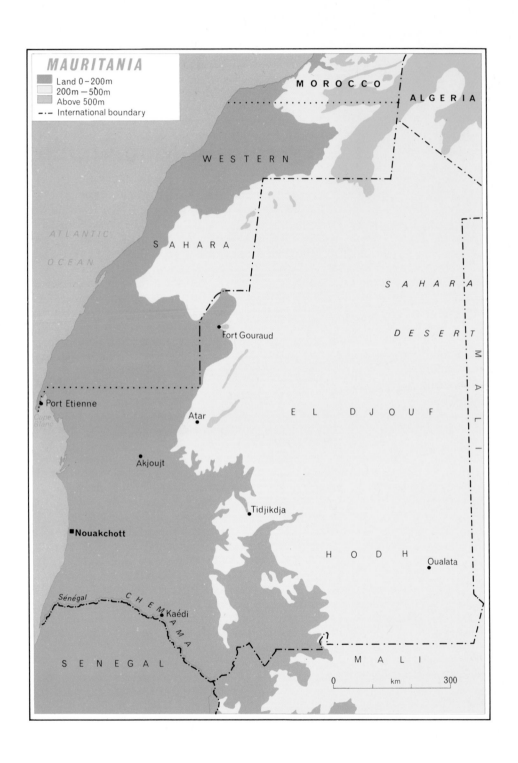

MAURITANIA

Land 0–200m
200m–500m
Above 500m
International boundary

MOROCCO   ALGERIA

WESTERN

ATLANTIC

OCEAN

SAHARA

SAHARA

DESERT

MALI

Fort Gouraud

EL   DJOUF

Port Etienne

Cape
Blanc

Atar

Akjoujt

Tidjikdja

HODH

Oualata

Nouakchott

Sénégal   CHEMAMA   Kaédi

SENEGAL   MALI

0   km   300

The Islamic Republic of Mauritania

The Islamic Republic of Mauritania is the most westerly as well as one of the most recent members of the League of Arab States, and one of the least known. Bounded by the Atlantic Ocean and the Western Sahara to the west, Senegal to the south, Mali to the south and east, and Algeria to the north-east, the area of Mauritania is approximately 1,085,000 square kilometres. Most of the north and east of the country forms part of the Sahara and is very sparsely populated. It is intrinsically a Saharan state, currently stricken by a fierce drought (most severe on the south-eastern edge), which has lasted for eight years, threatening not only the Tuareg nomads but all livestock and vegetation in its belt. The herds may one day be built up, the trees grown again, and the desert once more be peopled by nomads whose way of life it is; but at present there is only a small grain of hope unless the rains return. Most of Mauritania has an average rainfall of less than 200 mm a year. Nouakchott receives its water from a desalination plant.

The estimated population of Mauritania is 1·2 million, with large areas devoid of people.

The aridity of this country is due to its geographical position, lying within the tropics where desert conditions are paramount because of anti-cyclones. Therefore sand, aridity, and wind dominate.

Because of the desert heat, no river survives, and only the intermittent seasonal rivers in the regions of the plateaux give enough moisture for the palm-tree to grow. One river, however, the Senegal, which forms Mauritania's southern border, is free-flowing, but this is shared with her neighbours, Senegal and Mali. Along this river is a large cultivated belt known as the Chemama.

Temperatures in most of Mauritania are typically Saharan, but near the coast mean monthly minima range from 12°C to 21°C, and maxima from 26°C to 30°C. The coastline consists of sand-dunes and temporary lake-like marshes called *sebkhas* (often associated with salt deposits). The breezes from the Atlantic keep the temperature relatively cool, and it was partly for this reason that Nouakchott ('place of wind') was chosen as the country's capital. The old capital, St Louis, had to be replaced because it is in Senegal. Nouakchott was chosen also with an eye to future industrial expansion of minerals in the north, in the hope that these would provide the necessary capital for agricultural improvements.

Saharan Mauritania is basically the habitat of the dromedary and the nomad. Sixty per cent lies in the relentless pattern of a Saharan zone. Reafforestation programmes instigated by the government in Nouakchott are constantly in peril and at the mercy of privately owned herds.

Two-fifths of Mauritania is covered by sand of different colours and textures according to area: dunes which are fixed are generally coarse in texture and fawn in colour, while live dunes are crescent-shaped and formed of reddish wind-borne sand. The *erg* of the Trarza and Brakna territories is to be found in the 'fixed' vegetation of the Sahel.

Over 80 per cent of the people are pastoral nomads, wandering with their herds in search of the rare and scattered watering-places and pastures. Most of the remainder are engaged in settled agriculture in the south, although an increasing number are becoming urbanised owing to the remarkable recent industrial development of the mineral resources of the country.

Port Etienne, besides being a deep-water harbour, is also a good base for off-shore fishing, with access to a coastal shelf 65 kilometres wide abounding in fish, over which flows the plankton-rich cool Canary current. The fishing industry gives employment to a large number of people in that area.

RECENT HISTORY

Mauritania derives her name from the Moors—'land of the wandering Moors'—and is an apt description of most of the country. Both names are of Latin origin, and approximately 73 per cent of the people are Moors.

Mauritania became a French colony in 1904, and formed part of the former French West Africa. In 1957 she refused to join the Organisation Commune des Régions Sahariennes (O.C.R.S.) created by the French Government, but opted to become a self-governing member of the Communauté. In 1960 she declared herself an independent Islamic Republic, and in 1973 she became a member of the League of Arab States. This was a logical step in Mauritania's historical, social and political development. For centuries Arab culture has been a predominant influence in Mauritania, of particular significance during the days of the great flourishing of trans-Saharan trade. Arabic is her people's first or second language and the population is almost wholly Muslim. Thus, after French political domination had ceased, and particularly after the exploitation of her considerable mineral resources began in the early 1960s, thus increasingly freeing her from foreign economic domination too, it was not strange that Mauritania should become politically as well as culturally a part of the Arab World.

AGRICULTURE

The flocks, droves and herds of the pastoral nomadic population are vast: there are estimated to be 6,500,000 sheep and goats, almost 2,000,000 cattle, 700,000 camels and 300,000 donkeys in the country; the Hodh area of the south-west, with scattered oases such as Atar, is the most productive single area in this respect. For practical purposes nomadism as a type is determined by the amount of water available and the distance between wells:

    camel nomadism when distances are greater between watering points;

    goat and sheep nomadism when closeness of watering points is paramount;

    herding of cows in areas where water is more plentiful.

Livestock used to be the major export of Mauritania, with gum arabic and dates,

The nomadic people of Mauritania.

Drought has plagued the land for over eight years. The Sahara has devoured farm land once productive, and the camel must forage further and further afield. The cutting of the brush has left the path wide open for the desert to continue to move in.

The Land.

and it still plays a considerable part in the economic life of the country, despite her rapid industrial development.

In the Chemama area, on the northern side of the River Senegal, seasonal flooding has made the settled cultivation of a wide variety of crops possible. Millet and sorghum are the most important of these, but rice, maize, dates, henna, tobacco, melons and vegetables are also grown. Gum arabic is still marketed from the acacia thorn trees. The area provides seasonal pasture for cattle; and simple earth dams have greatly improved the control of the waters of the Senegal. *Marigots* (side channels) are phenomena which are of particular importance to southern Mauritania. These are a network of stagnant branches which feed back into the River Senegal during the dry season. The stagnant reaches 'capture' many fish while the side channels are an irrigation boon, forming natural irrigation canals. As the level of the Senegal increases to a maximum in September each year, so do the *marigots*, such as Gorgol Noir and Gorgol Blanc. Settled agriculture is also practised in these Gorgol valleys and in isolated patches of ground where well-water is available; for example, in the plains of Trarza and Brakna.

On the boundary with Rio de Oro there is more moisture, which in turn leads to the support of vegetation such as acacias, shrubs and grasses; this is a camel-rearing area. Farther north the climate becomes drier and the dunes more shifting.

Although the role of agriculture in Mauritania has in the past few years been somewhat overshadowed by industrial development, it remains a vital constituent of the country's economy, and her growing self-sufficiency and prosperity will enable more ambitious irrigation and land-reclamation projects to be carried out in the future.

INDUSTRY

Mauritania is being transformed economically by the discovery of 250 million tons of high-grade haematite iron ore in the Kedia d'Idjil hills in the north-west, and the construction of a railway from the workings at Fderik (Fort Gouraud) to the coast at Nouadhibou (Port Etienne). Whereas previously considerable foreign aid had been required to support Mauritania's economy, since 1960 her gross domestic product has increased by 10 per cent per annum, she has a favourable balance of trade, and the prospects look very encouraging. In addition, a 30-million ton deposit of copper has been discovered at Akjoujt, while prospecting is taking place for oil, gypsum, titanium and manganese. Foreign investment in all these industrial activities is considerable, and the improvement of communications, transport and urban development is rapid, not to mention the growth of health, education and other social facilities. Mauritania's future appears to be that of steady economic improvement, and she will undoubtedly play a significant role in the affairs of the Arab World, of which she is now formally as well as historically an integral part.

REFERENCES

Alfred G. Gerteiny, *Mauritania*, The Pall Mall Press, 1967 (Pall Mall Library of African Affairs)
National Geographic Magazine, Vol. 145, No. 4, April 1974, National Geographic Society, Washington D.C.

The Republic of Lebanon

Lebanon is one of the smallest independent republics, with an area of only about 10,400 square kilometres and a population of 2,790,000. Compared with the larger Arab countries, it is densely peopled, with an average of nearly 945 inhabitants per square kilometre of arable land. The high birth rate of 2·5 per cent a year is worsening the situation. The country has a coastline of some 200 kilometres along the eastern shore of the Mediterranean, but extends inland for only 50 to 80 kilometres, and from this narrow, broken coastal strip, the Lebanon Mountains really dominate the country. They run parallel with the coast, rising quite steeply through a series of foothills to reach a height of over 3,000 metres in the north, and rather less in the south. East of this range the land falls abruptly into a broad lowland valley some 120 kilometres long and 16 to 20 kilometres wide, drained by the Litani River in the south and the Orontes in the north. It is known as the Beqaa, and for the most part is at an altitude of between 600 and 900 metres. Still farther east the land rises steeply again from the eastern floor of the Beqaa to become the Anti-Lebanon Range, along the crest of which runs the frontier with Syria, and which, because of steep slopes, poor soil, and scarcity of water in summer, provides only nomadic grazing.

Lebanon is basically an agricultural country, more than half its population living in rural areas and depending for their livelihood on some form of agricultural activity. But as a result of the great expansion and development of Lebanon's mercantile economy, although the main occupation is agrarian the economy is not an agricultural one, less than one-fifth of the national income being earned from this source. The economy is predominantly supported by the commercial sector in which the city of Beirut, with its important harbour and entrepôt trade, its highly-developed banking services and its attraction as a tourist centre, is prominent. The Lebanese are great traders: they sell their own fruits, vegetables, and textiles to their Arab neighbours at highly competitive prices. The average earnings of an individual in a commercial occupation are appreciably higher than in agriculture.

At present, agricultural crops are produced from about 38 per cent of the whole country. Only 75,000 hectares are irrigated by river water, the rest depending on rain for its water-supply, leading to far more intensive cultivation and the consequent terracing of the accessible and well-watered slopes of the mountains. In comparison with most other areas of the Middle East, Lebanon is well endowed with water, particularly in the winter season. But of the total area of cultivable land, about 40 per cent of the whole, less than half is actually being cultivated. This is

The development of agriculture is influenced by the high mountains and the comparatively plentiful supply of spring water. Because of the differing levels of the topography, a wide range of crops can be grown, with olives, vines and figs on the lowest foothills.

Harnessing the Litani River where the flow is at its maximum in the winter months and at its lowest in the summer months.

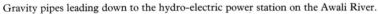

Gravity pipes leading down to the hydro-electric power station on the Awali River.

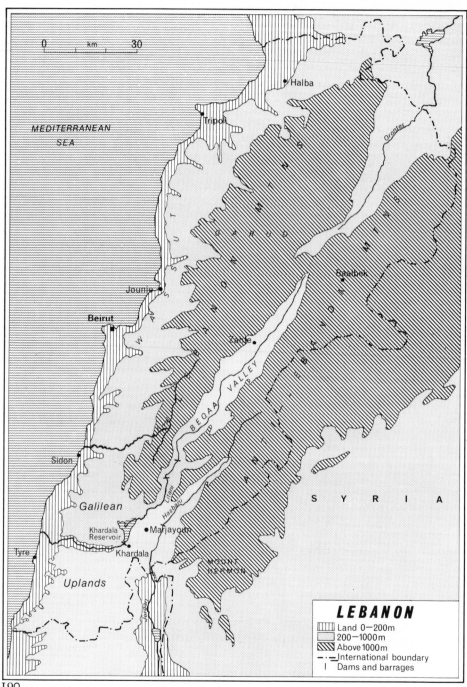

MEDITERRANEAN
SEA

•Halba

•Tripoli

Orontes

A N T I L E B A N O N   M T N S

•Baalbek

Jounie•

Beirut•

Zahle•

B E Q A A   V A L L E Y

A N T I L E B A N O N

Sidon•

S Y R I A

Galilean

Litani

Hasbani

Khardala
Reservoir

•Marjayoun

Tyre•

•Khardala

MOUNT
HERMON

Uplands

Jordan

LEBANON

⬚	Land 0—200m
⬚	200—1000m
⬚	Above 1000m
–·–	International boundary
I	Dams and barrages

mainly because of the high cost of working the less productive land, which there-fore tends to lie fallow. Three hundred thousand hectares can be used for cultiva-tion. The pastures, non-arable lands, are overstocked, mainly with sheep. The greater amount of precipitation has a lot to do with this, as the soil tends to be fertile even at the highest elevation.

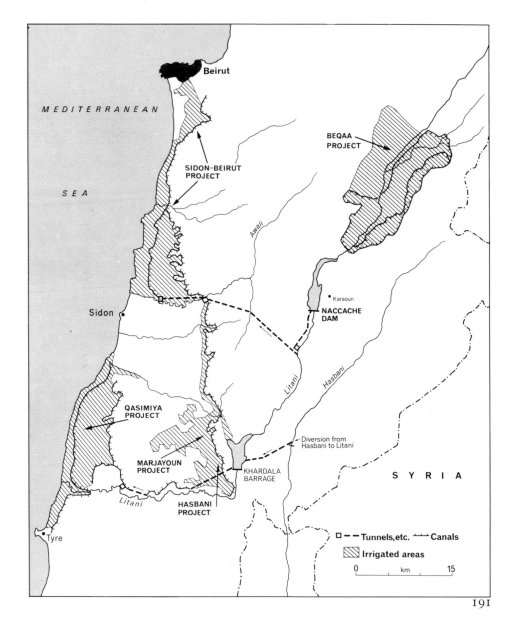

DEVELOPING AGRICULTURE
Lebanon can be divided into three major agricultural regions: the coastal plains, the Beqaa Valley and the mountainous region.

The Coastal Plains

For most of their length these form a very narrow strip of flat land which varies in width between 3 and 22 kilometres, broadening out in the north round Tripoli and Halba, and in the south at Tyre and Sidon. This region is agriculturally the most productive in Lebanon. It enjoys a Mediterranean type of climate, with moderate, wet winters during which occur over 1,000 mm of annual rainfall on the windward slopes of the Lebanon Mountains, and about 1,270 mm even higher up; and with rainless summers that are warm and uncomfortably humid, with average temperatures ranging from 14°C in January to 28°C in August. The need for water during the growing season is met by irrigation, using water from the many streams which are fed by mountain springs and flow down to the sea. Because of the high fertility and the abundant water-supply, more than one crop can be grown on the same land during the year, and this is so especially in the case of pulses and vegetables. These plains are also the main area for the growing of citrus fruits and bananas. All these products are regarded as cash crops and the greater part of their production is exported. Most of the agricultural development schemes in Lebanon are to be found in this coastal belt, where it is relatively easy to control the rivers and thereby extend the blessing of irrigation to land hitherto uncultivated.

The coastal plains widen out in the north, but here the lack of water for irrigation and the damage done to crops by very strong cold winds result in greatly reduced productivity.

The Beqaa Valley

This deep, fertile lowland trough at an altitude of 600–900 metres between the Lebanon and Anti-Lebanon Ranges is part of the Rift Valley which extends to the Red Sea, and is drained by two important rivers, the Orontes flowing northwards into Syria, and the Litani flowing southwards along the whole length of the valley until its waters finally turn westwards and enter the Mediterranean near Tyre. This valley-plain lies in the rain shadow of these mountains and forms the largest continuous agricultural tract in the country. Besides being the principal intensive cereal-growing area in Lebanon, in rotation with sugar beet, potatoes, and fodder crops, it produces some cotton and a wide variety of food and cash crops including vegetables, fruit (particularly citrus and bananas) and, especially in recent years, tobacco, onions and garlic. The introduction of the sunflower for its important oil is a recent development.

Certainly the agricultural development of the Beqaa is being intensively pursued. While towards the hills and particularly near Zahle there are market gardens, vineyards and orchards, the northern part is of limited productivity because of lack of rainfall, thinner and poorer soil, and the steepness of the terrain. The flatter central and southern parts, however, are being fully developed by applying the most

Close to the rich lands of the Beqaa are the harsh and infertile eastern foothills of the mountains. Only primitive wooden ploughs attached to a yoke can, as yet, be used in the cultivation of this difficult terrain.

modern methods of production, including widespread mechanisation, the most up-to-date means of irrigation, the increased use of fertilisers and the improvement of means of transporting agricultural products to ports and local markets. Even so, there remains room for further agricultural expansion in the Beqaa. There are a number of state-run irrigation projects among which the most important, the Litani project, will help to irrigate a large area of some 26,000 hectares in the Beqaa plain, and in south Lebanon, where rocky outcrops and marshland are difficult for agriculture.

The Mountainous Region

This comprises the main chain of the Lebanon Mountains, of sandstone and limestone deeply dissected with dramatic gorges such as that made by the Litani River 270 metres in depth, and the belt of foothills on the seaward side. Cultivation is mostly confined to the lower western slopes, where abundant water and good soil are particularly advantageous. It is here that olives, vines and soft fruits are grown on fertile terraces carved in the steep hillsides with intense human effort, while wheat is grown wherever the land is level enough. The mountains provide the source of a great many underground springs, which ultimately produce the much needed water for irrigation, but only in a limited area which grows peas, beans, lentils, and other vegetables. Future development is going hand in hand with the increase in the demand for fruit and vegetables from the towns, and citrus fruit groves are on the increase. Bananas do well. The Lebanese tend to use the method of interplanting bananas with citrus trees, as the former bear fruit four years earlier, the latter needing six or seven years to come to fruition.

The highland area above 900 metres once contained dense forests of cedar, pine and oak, since cleared, and is now given over to apples, pears, cherries, peaches and other fruit. However, a special programme is being implemented of reafforestation, not only because of the great demand for timber, but as large-scale prevention of soil erosion, in which the roots of trees play a vital role.

193

Scenic beauty of the western hill country with its pine stands.

Laban is Aramaic for white. In the area above Tripoli, snow-capped mountains form a background to some of the few remaining cedars.

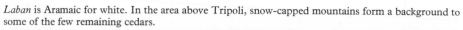

Cultivated valley, showing the ancient skill of terracing even the steeper slopes.

Banana plants thrive on hot, damp coastal land, and since they bear fruit at an early stage they can economically be interplanted with citrus trees.

OWNERSHIP AND ECONOMIC USE OF LAND

The large farms in Lebanon are mainly on the coast or in the fertile plains of the interior. One reason for this is that much of the terrain is steep and can be cultivated only if a great deal of labour is spent in terracing it; nevertheless the Government has allotted money for terracing several hundred slopes, deep ploughing, and the construction of irrigation canals. The steady movement of the rural population to the towns and, indeed, to foreign countries has helped to some extent to lessen the concentration of economic and political power into the hands of a landowning class. About 200 landlords own over half the cultivated land; the rest of the land is in 'parcelised' holdings of from 2 to 100 hectares, and owned by comparatively few farmers.

The economic use of agricultural land in Lebanon is similar to that in Syria, about 1 per cent paying rent in cash or kind. Farmers are helped by the Agricultural, Industrial, and Real Estate Credit Bank, and generally repay the Government in annual instalments over a period of ten years.

AGRICULTURAL PRODUCTION AND ITS POTENTIALITIES

Being blessed with good topography, varied soils and differing climates, Lebanon produces a great variety of crops, but it is worth noting that 85 per cent of the total income from agriculture comes from citrus fruit, bananas, sugar cane, tobacco, olives, grapes, apples, pulses, vegetables, and mixed farming, while the remaining 15 per cent comes from timber and animal products. Development plans to raise the quality and amount particularly of poultry are being most successful. Farmers are encouraged to produce more forage crops, by price protection for local corn and sorghum, by the introduction of high-yielding cereals, and by the better use of waste products such as animal manure, banana leaves and stalks, carrot and apple pulp, and the wood shavings used for poultry litter.

Until 1962, development planning hardly existed. Before 1958 Lebanon had no clearly-defined policy for economic development. What projects there were came into being when the State was confronted by some emergency or wished to achieve some political objective. There was no planning, no co-ordination of different projects; in fact, no nationwide design for the future use of resources.

Early in 1962 the Government drew up a five-year economic development plan. Although this plan acknowledged the need for the State to intervene in development, it was based on the theory of a free economy, namely that economic development can be achieved by the inherent forces of a free market, and that the public sector should only invest in certain limited fields where it will not infringe upon the rights of private enterprise. The Government's 'Green Plan', which aims at bringing 100,000 hectares of arid land under cultivation, is being slowly implemented.

The Use of Traditional Methods

Many Lebanese farmers still cultivate their land by primitive and traditional methods and equipment: the plough is still of wood and is pulled by animals; the harvest is reaped by hand; the system of crop rotation is unscientific; the use of fertilisers is limited to relatively few crops. The small average size of the farmer's holding stands in the way of his modernising his methods and prevents him from obtaining a proper economic return from his land. Yet by processing agro-industrial by-products and the reclamation of one-quarter of the whole area, un-cultivated lands could be used to augment the Lebanese farmer's economic difficulties, and encourage him so that the nutrient needs of all livestock could be met with a surplus of feed available to support large numbers of, say, dairy cattle, somewhat on the lines of agricultural co-operatives if necessary.

Finance

Most Lebanese farmers are, as we have seen, small operators whose purchasing power is too low to permit them to invest in short- or long-term improvements; for instance, apples require specialised refrigeration, storing, etc., and for this capital is needed. This type of undertaking is out of reach of the small farmer. However, encouragement given by the Government's current Development Plan is bearing fruit. Already there is a great improvement in agricultural production: a change of methods in shifting to products of high value, such as those already mentioned, larger irrigation areas, the better use of fertilisers and pesticides, crop rotation and improved varieties of seeds. The increase in the apple crop, for instance, is typical of this type of new planning; with careful analysis of soils to suit the type of apple trees planted, a high yield is obtained.

Expanding Markets

As the home market in Lebanon expands with the growth in population, a considerable proportion of its agricultural production is used for domestic consumption. However, with competitive prices, modernisation in sorting, grading, packing and despatch, as well as refrigeration, Lebanese products compete favourably with those of other nations, even in neighbouring Arab countries, and exports are expected to grow in the coming years.

Scarcity of Water Resources

We have seen that most crops in Lebanon depend on rainfall for their water. Although there are ten sizeable rivers in the country, the mountainous nature of the topography does not allow the fullest use of their waters without a comprehensive plan for the rational use of water for irrigation.

The Government has initiated a detailed survey of existing facilities and a number of irrigation projects, both of the surface and underground water supply. Among those completed are the Qasimiya project for the irrigation of nearly 4,000 hectares and the Marjayoun project which brings water to some 10,000 hectares. In 1960 the total area irrigated amounted to 67,000 hectares, of which 77 per cent was put down to annual crops with the remainder in permanent crops. The Government also aims to make better use of the Litani and Hasbani rivers. The Litani for much of its course flows at a high altitude where its waters can be contained and stored for maximum return from irrigation and electricity.

The Litani River is about 169 kilometres long and it discharges about 755 million cubic metres of water per year. In 1955 the project for making greater use of both the Litani and Hasbani rivers by harnessing the waters for irrigation and for generating electric power was drawn up and is being implemented in stages, the last of which is due to be completed in 1977. A dam 1,000 metres wide with a storage capacity of 60 million cubic metres of water has already been built near the village of Pharaon to irrigate the lands of the southern Beqaa and to bring irrigated cultivation, such as forage crops for cattle and sheep, to the hitherto unwatered area between Beirut and Sidon.

The Hasbani River projects will harness the waters of this stream at a point only 16 kilometres from Lebanon's frontier by means of a dam which will make it possible to store six million cubic metres of water and to divert a further 50 million cubic metres to the reservoir to be built at Khardala, on the Litani River. Thus, of the 156 million cubic metres of water which at present flows across the frontier, 110 million cubic metres will be used by Lebanon to irrigate over 15,000 hectares in the river valley and in the plain of Tyre, besides generating 20,000 kWh of electric power.

This river, the Hasbani, as a tributary of the Jordan River, is a very important water resource which directly affects not only Lebanon, but Jordan, Palestine, and Syria as well.

The Naccache Dam at Karaoun on the Litani River.

THE LITANI DEVELOPMENT

The Litani River basin, by the nature of its water supply, could be described as illustrative of many of the problems encountered in the development of water resources in the Middle East. (Obviously, each river basin is in itself unique.) Unlike the Nile, Jordan, Tigris and Euphrates basins, the Litani basin lies almost entirely within the Lebanon. About 80 square kilometres of Syrian territory could be considered within the basin and might contribute an unsubstantial amount of winter flood water to the Litani.

In the Litani basin water run-off is higher than average, as the greatest areas of precipitation are the mountainous areas with steep, bare slopes, subject to rainfall and melted snow. Also water passes through the permeable limestones of the mountains and then reaches the rivers via springs.

Key to the Litani development was a storage dam sited at the southern end of the Beqaa near Karaoun, with two hydro-electric systems; one continued down the Litani from the dam, while the other, fed by a tunnel from the reservoir through the Lebanese Mountains, continued down the Awali River. In this way the two systems utilise both the inflow of the Litani below the dam, and the flow of the Awali (see map) estimated at over 100 milliard cubic metres annually.

Irrigation could then be brought to part of the southern Beqaa, areas of better land in the Galilean Uplands, and parts of the Sidon–Beirut area near the coast. And in this way there would be sufficient water for the Qasimiya Project on the coast to meet peak summer demands.

The site chosen for the principal dam and storage reservoir was near the village of Karaoun, at an altitude of 795 metres above sea level, just before the Litani begins its descent from the high plain into a deep gorge. The Naccache Dam at Karaoun, more than 60 metres high, created a reservoir which now retains 220 million cubic metres of water.

From this reservoir the Markabe Tunnel, which lies at the same altitude as the dam over its 6-kilometre length, carries a proportion of the water to a point on the steep-sided mountains downstream (above the dry river bed). The water from the tunnel then flows abruptly for 180 metres through an almost vertical tube within the mountain, and continues into the underground Abdel-Aal Power Plant. From this plant the water is then recycled; a small dam diverts it from the plant and uses a major spring which enters the river at this juncture and, via a tunnel, the Awali. It is an inspiring engineering feat which enables the water to travel for 16 kilometres beneath the Lebanon Mountains before emerging on the seaward side. The water emerges from the tunnel at an altitude of 600 metres into a small lake at Kanane in the well-wooded hills above Sidon.

Construction of water projects is always subject to 'something going wrong'— for instance, the tunnel under construction was blocked when a section containing water under high pressure was reached, which produced flooding and cave-ins. This increased the cost, and work was not resumed for about two years.

Another power plant downstream at Jounie, 210 metres above sea level, carries the water along the coast for irrigation purposes (see map).

A more recent project, embarked on in 1974, involves the straightening of the channel of the River Bardouni at Zahle. This will make yet a further contribution towards the harnessing of water for the agricultural needs of the Lebanese people, and thus aid that type of development on which the future prosperity of the country so much depends.

REFERENCES

W. B. Fisher, *The Middle East*, Methuen, 1971

Peter Mansfield, *The Middle East : A Political and Economic Survey*, Oxford University Press

Greene, Brook, and Farhat Hafiz, *The Feed-Livestock Economy of Lebanon with Projection to 1976 and 1981*, American University of Beirut, Lebanon 1973

Alice Taylor (Ed.), *The Middle East*, published in co-operation with the American Geographical Society, David and Charles, 1972

L'Economie Rurale Libanaise, No. 41, 1972

The Middle East Journal, Winter 1971, Middle East Institute, Washington D.C.

Recueil Statistiques Libanaises, No. 8, Année 1972, Le Ministère du Plan Direction Centrale de la Statistique, Beyrouth, Liban

The Syrian Arab Republic

The Arab Republic of Syria lies on the eastern shore of the Mediterranean, along which it has a coastline of about 160 kilometres. The country lies to the south of Turkey, to the north and east of Lebanon, to the west of Iraq and to the north of Jordan. The frontiers of Syria are largely artificial (for example, its northern boundary for the most part follows the Baghdad Railway), but they do coincide with such natural features as the Mediterranean Sea and the crest of the Anti-Lebanon in the west, and even with a few kilometres of the upper reaches of the River Tigris in the extreme north-east.

Syria falls geographically into two broad natural regions: the narrow western zone consisting of a coastal plain with a parallel system of mountain ranges and valleys, it being in this zone that the majority of the population live; and a much broader, vaster, eastern zone consisting of an arid upland plateau covered with steppe in the west but becoming a desert farther east and sloping gradually down, first to the valley of the Euphrates which crosses the country and then, in the extreme north-east, to that of the Tigris. In the south-west Damascus, the capital, lies in a basin partly formed by the Anti-Lebanon in the west and the Jebel Druze volcanic uplift in the south. Temperatures in Damascus average 23°C in summer and 10°C in winter, with between 20 and 52 per cent humidity.

Much of Syria receives under 250 mm of rainfall per year, and this becomes less and less towards the east; but in the western zone near the Mediterranean the annual rainfall is more abundant and is as high as 1,000 mm in the mountainous coastal belt. There is a narrow, curving arc of moderate annual rainfall, about 250 mm, known historically as the 'Fertile Crescent'. From its southern point in Jordan it spreads north into Syria, along the eastern foothills of the Anti-Lebanon, and widens to embrace a line of fertile basins as far as that of Aleppo. It then turns eastwards along the northern border with Turkey, into the Gezira, and continues along the valley of the Euphrates, whence it gradually thins out as the rainfall becomes inadequate.

Syria has a total area of about 185,180 square kilometres, of which only approximately 45 per cent is suitable for arable cultivation, while the remainder of the country consists of bare, mountainous deserts and some forests, which support a form of grazing only suitable for nomads. At the 1970 census the total population was 6,924,000, giving an average growth rate of 3·3 per cent, which is high when compared with most Arab countries. What is so marked in Syria is the tremendous contrast between what might be termed the settled region, which contains not only 95 per cent of the population but the total cultivable regions as well, and the vast

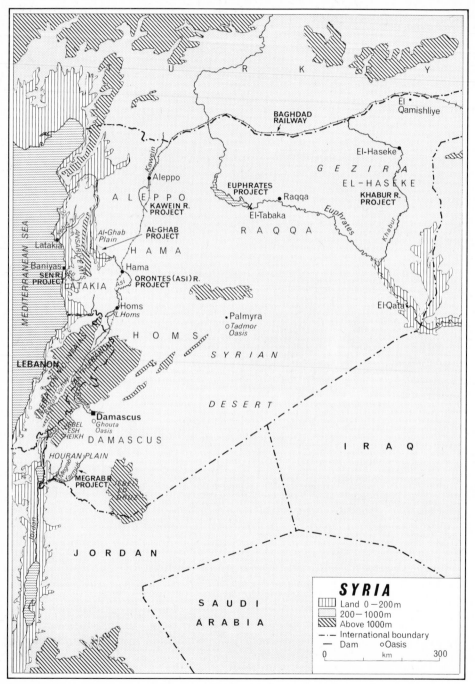

T U R K E Y

BAGHDAD
RAILWAY

El
Qamishliye

El-Haseke

G E Z I R A

EL-HASEKE

EUPHRATES
PROJECT

Raqqa

KHABUR R.
PROJECT

El-Tabaka

Aleppo

A L E P P O

KAWEIN R.
PROJECT

R A Q Q A

Al-Ghab
Plain

AL-GHAB
PROJECT

H A M A

Latakia

Hama

ORONTES (ASI) R.
PROJECT

El-Qata

SENRJ
PROJECT

Baniyas

MEDITERRANEAN SEA

Asi

LATAKIA

Homs
L. Homs

H O M S

Palmyra

Tadmor
Oasis

S Y R I A N

LEBANON

D E S E R T

Damascus
Ghouta
Oasis

I R A Q

JEBEL
ESH
SHEIKH

D A M A S C U S

HOURAN PLAIN

MEGRAB R.
PROJECT

Jordan

J O R D A N

SAUDI

ARABIA

SYRIA

|||| Land 0 – 200 m
200 – 1000 m
Above 1000 m
– · – · International boundary
— Dam   o Oasis

0   km   300

202

uninhabited areas in the central and eastern zones of the country, which are so arid as to be able to support only nomads, mainly Bedouin. More than half of the cultivable land is in the provinces of Aleppo, Raqqa and El-Haseke. In 1970, out of a potentially cultivable area of 8·7 million hectares, only 32 per cent of the total area was actually cultivated.

The reason why so large an acreage in Syria is not cultivated is the shortage of water. Water is scarce because much of the country lies in a rain-shadow, being bordered on the north and west by mountain chains which effectively prevent wind-borne rain from reaching it. The only exception is the Fertile Crescent, where rainfall is adequate to support cultivation without recourse to irrigation. Although the Syrian part of this Crescent is rain-fed, intensive cultivation is in general confined to such areas as the Houran Plain, the Damascus oasis, the districts of Homs and Hama, and Aleppo.

Fortunately, to compensate for her scanty rainfall, Syria has a number of rivers. The two most important are the Euphrates and its tributary the Khabur, which together contribute about 90 per cent of the total discharge of all the rivers in Syria. The Orontes river (also called the Asi), the largest in western Syria, rises in the Lebanon and flows north along the Beqaa Valley into Syria and then through a once swampy basin to cross into Turkish territory just before entering the Mediterranean. With the regulating of the Orontes, such swamps have mostly been reclaimed for agriculture. The Barada river rises in the Anti-Lebanon, flows southeast and irrigates the important Ghouta oasis near Damascus. In the extreme south the Yarmuk river rises in Syria and flows south-west along the frontier to become the most important tributary of the River Jordan. Although the flow of these rivers is to a large extent seasonal, the control of their waters is effectively contributing to the future development and prosperity of Syrian agriculture. At present the larger percentage of the cultivated area is dependent entirely on rain for its water-supply.

RAINFALL AND AGRICULTURE

As we have already noted, rainfall is the controlling factor in Syrian agriculture. Two major crops, wheat and barley, are dependent on this, however low it may be. On the other hand, cotton (medium staple) has been developed over the years in the provinces of Hama and Homs as an important cash crop, primarily because it is less dependent on rainfall. The Ministry of Agriculture and Agrarian Reform has geared its policy towards increased output by giving more impetus to farmers in the form of loans, fertilisers, better seed, and technical aid (advice and mechanisation). This applies equally to the cultivation of tobacco, which receives special government support. Sugar beet was introduced fairly early in the 1950s and is doing well.

The country may usefully be divided into five rainfall areas:

1. Regions where rainfall exceeds 1,000 mm annually

These extend over an area of almost 400,000 hectares in the Ansariye Mountains in Latakia province, and in the Jebel esh-Sheikh in the extreme south-west of Damascus province. These mountains are covered with forests and the amount of rain they get is sufficient for agricultural development if the forests are cleared. Much of the winter precipitation is in the form of snow.

2. Regions where annual rainfall varies between 600 and 1,000 mm

These total some 1·9 million hectares and lie on either side of the Ansariye mountain chain, in a small area south of the Jebel esh-Sheikh, and in the north-eastern tip of El-Haseke province. These areas produce winter grain, particularly wheat, barley and vegetables.

3. Regions where annual rainfall varies between 400 and 600 mm

This area is rather larger than the regions mentioned above; it extends over a narrow ribbon of land, varying in width between 30 and 50 kilometres, lying to the east of the Ansariye and Jebel esh-Sheikh regions and gradually broadening out into the high Aleppo area. There is much the same amount of rainfall also in the far north-eastern tip of the country where the influence of the nearby Zagros Mountains of eastern Turkey begins to be felt. Agricultural expansion has taken place in these regions during the past two decades and in a good year they now contribute 40 per cent of the wheat crop. Unfortunately, great fluctuations in the rainfall from year to year considerably affect production.

In the province of El-Haseke, the increasing expansion of the Gezira, which is rain-fed, after the Second World War, from 20,000 hectares to today's total of 1·6 million hectares has, as a natural consequence, resulted in an upsurge of population concentration. The main town of El-Qamishliye, which acts as a commercial centre, is already seventh in order of city or urban population in Syria. Nevertheless, with the provinces of Raqqa and Deir ez-Zor, these areas compose more than 40 per cent of the total land area with only 15 per cent of the population.

4. Regions where the annual rainfall varies between 200 and 400 mm

These are semi-arid areas of steppeland and total some six million hectares. They are distributed among all the Syrian provinces with the exception of the wetter Latakia province. Dry farming is prevalent in some of these areas, but harvests are to a large extent uncertain except where small-scale irrigation schemes and the installation of pumps help to bring the water from underground artesian sources.

5. Regions where annual rainfall varies between 100 and 200 mm

This is the largest area of all, covering over 80 million square kilometres or 40 per cent of the total area of Syria. Here the amount of rainfall is really insufficient to support dry agriculture but it is enough for the growth of grass, which renders the Syrian desert good pastureland in winter and spring. Dry farming had nevertheless been expanded to its limit in the wetter parts of this area before further expansion of the cultivated area was achieved with the help of irrigation.

The ancient *norias* (*sakias*) of Hama are still turning, still providing water for irrigation.

IRRIGATION PROJECTS

Most of the unworked cultivable land lies in the region where annual rainfall varies from 400 to 600 mm and in the semi-arid areas where it varies from 200 to 400 mm. This land remains unworked and unexploited largely because it lacks water. Thus it is abundantly clear that irrigation projects are vital to the further development of Syria's agriculture.

The area of land at present under irrigation by water from rivers, wells and springs is about 670,000 hectares. Preliminary studies indicate that Syria's present water capacity cannot provide irrigation for more than 1½ million hectares unless the various proposed irrigation projects on the Euphrates, Khabur, Orontes and other rivers are put into effect.

In the past irrigation was confined to the age-old primitive systems such as *norias*, and all was small-scale. After the 40s the government gave great attention to land improvement schemes with irrigation, mechanisation and large-scale projects given priority and budgeted for.

The extent of irrigated land has thus increased considerably since the Second World War, due to the efforts of farmers on the one hand and to the implementation of government projects on the other. The Euphrates Dam is the most ambitious of all Syria's irrigation projects in her five-year development plans.

In 1973 at a special ceremony the end of the first stage of the main structure of the dam was marked when the course of the river was diverted. The dam's first turbine came into operation in 1974. Farmers are now making greater use of pumping-machinery to raise the water of the Euphrates and its tributary the Khabur, and from wells in the provinces of Hama and Homs. The result has been that the area irrigated by pumps has increased to over 300,000 hectares.

The government's contribution to the increase in the irrigated area took the form of a number of projects by which water was brought to irrigate some 50,000 hectares for the first time (see map).

The most important of these projects are:

The Orontes river project (22,000 hectares) in Hama and Homs;

The Sen river project (4,000 hectares) on the coast near Baniyas;

The Khabur river project (5,000 hectares) in the Gezira region;

The Kawein river project (16,000 hectares) south of Aleppo;

The Megrab river project (2,000 hectares); this river is a tributary of the River Yarmuk.

Last but certainly far from least, the most important scheme completed by the Government is the al-Ghab project on the Orontes river north of Hama. The Ghab, being a lowland trough not more than 15 kilometres wide, has a climatic regime which is quite distinctive from that of the highlands (relief plays such an important part in the control of rainfall). This region has a similarity to a steppe climate, with moderate to low rainfall, cold winters, and hot summers. This scheme serves a number of purposes including irrigation, drainage, and electricity generation. It involved the construction of two major barrages, which not only 'control' the Orontes but eliminated the marshland, and thus furthered irrigation with consequent resettlement and in turn the production of a variety of summer and winter crops in this fertile soil. Wheat, barley, cotton, rice, and sugar beet are all grown here. This has been achieved in an area which was formerly quite unproductive. The scheme also includes land reclamation in the al-Ashar and al-Ghab plains. These two plains have an area of 28,000 and 50,000 hectares respectively, giving a total of 78,000 hectares, of which 35,000 were once covered by swamps that have now been completely drained. Dams had to be built at al-Rastin, Mahreda and al-Asharna. The course of the Orontes river was deepened and widened between the villages of Kanwi and el-Kafir, and the basalt promontory, which prevented the discharge of water from the el-Kafir river, was cut through to enable the two rivers to unite. The project also required the digging of a network of irrigation channels in the al-Asharna region.

The Euphrates Dam

This scheme, which is regarded as a major factor in the development of the Syrian agricultural economy, was begun in 1968, after agreement had been reached between the two other countries who are equally concerned with retaining a share in the water of the Euphrates, namely Turkey and Iraq.

The area of the section of the Euphrates basin that lies in Syria is about 70,000 square kilometres, most of which is suitable for agriculture. The total annual discharge of the river in Syrian territory is 26 milliard cubic metres, which represents 80 per cent of Syria's total water resources. Studies had shown the feasibility of building a dam at Tabaka, 200 kilometres east of Aleppo, between Kalek and Negm and Raqqa. The third and final stage of this construction is due to be completed soon, when a reservoir will be formed by diverting the course of the river, which when filled will store 40,000 million cubic metres of water. This will provide some 600,000 hectares of land with valuable irrigation water. In addition the dam will facilitate the generation of hydro-electric power. Some of this power will be used to work water-pumps to irrigate the highland areas on both sides of the river, while the rest will be used for industrial purposes and for local domestic consumption in northern Syria.

With progress of this nature there are always the casualties: for instance, farmers whose lands and villages are to be submerged will have to be resettled. And then in 1969 there was a series of sudden and dangerous floods on the Euphrates, the most serious of which was on 24 January, when the river discharge reached 5,100 cubic metres per second. The major flood normally occurs in April, when it inundates all the surrounding land, but on this occasion, owing to the erection of protective flood barrages, work was able to continue on the foundations of the main dam and hydro-electric power station. Nevertheless, this 70-metre high dam will literally transform an area which hitherto could not be termed productive by modern standards into an agricultural area of major significance.

A transformer station has been set up in the Sheikh Said region in Aleppo province, and the 160-kilometre long 220-kw Kadra transmission line between Aleppo and al-Tabaka is in operation.

Pre-fabricated reinforced concrete pipes, 3 metres in diameter, are used for lining a tunnel from which the foundations of the dam can be monitored. The Gezira area with its large-scale cotton cultivation will benefit considerably when the Euphrates Dam is completed, and will increase in importance.

The Khabur River Project

This project aims mainly at diverting the waters of a section of the Khabur river by barrages in order to irrigate an area of fertile land on its left bank between Tel Manas and El-Haseke, in Gezira. An irrigation network for the distribution of the water, consisting of a main channel 68 kilometres long plus more than fifty branch channels with a total length of about 125 kilometres, has been completed. This complex system of drainage canals to dispose of salt deposits will result in an area of some 100,000 hectares becoming productive land which prior to this was little used.

The Yarmuk Project

The principal aim of this scheme is to use the waters of the springs in the Houran plains which slope down to the Yarmuk Valley, and thus to irrigate nearly 3,200 hectares in addition to the 2,000 hectares of existing irrigated land in the upper reaches. And in order to make maximum use of these waters two storage dams are being constructed with canal extensions to bring water to a greater area.

The Upper Orontes Project

This project requires the construction of barrages on the Orontes river close to the Syrian–Lebanese border. The water stored by the barrages will feed a number of channels and will irrigate an area of some 12,000 hectares on either side of the river mouth of Lake Homs.

The Diversion of the River Jordan and its Tributaries

One of the most important projects which must be given proper attention in any plans for land development in Syria is the diversion of the River Jordan and its tributaries. The use of its waters is of concern not only to Syria, which already has two storage dams (one under construction and the other planned) on the Yarmuk, but to the Lebanon, Jordan and Palestine as well, because it directly affects all of them. Additional information on this and other projects for the exploitation of this vital river is given in the section headed *The Waters of the Jordan* in the chapter on The Kingdom of Jordan (pp. 237–246). Syria attaches great importance to irrigation projects because agriculture is still the lifeline of the Syrian economy despite industrialisation in recent years and trading trends in policy, and particularly to the construction of the Euphrates Dam and other smaller dams which will, during the next fifteen years, add an area of some 450,000 hectares to the present irrigated land. Of this newly irrigated area, 93,000 hectares have been completed during the Third Five-Year Plan, 1971–75.

LAND OWNERSHIP

Studies in land ownership have revealed that the minimum size of agricultural unit which can bring a farmer and his family (an average of six people) a decent standard of living is about 10 hectares.

Before the land reform of 1958, large-scale ownership was predominant (30 per cent of the agricultural land was owned by only 1 per cent of the population), while smallholdings, usually directly owned and worked by individual farmers, accounted for less than 15 per cent of the total area of land. Even if we ignore the 23 per cent of state-owned lands, the percentage of smallholdings was still less than 21 per cent. This imbalance was aggravated by the incongruity between the size of the holdings and the number of owners, for 70 per cent of the rural population were landless, and lived as tenants or worked as hired labour.

In 1969 further much-needed agrarian reform was implemented. This limited land holdings to 80 hectares of land which was irrigated, or 300 hectares of rain-fed land per person. It was early on in 1975 that the first stage of land reform was completed: 708,000 hectares in 1,413 villages were re-allocated to 40,000 families.

Size of holding was governed by the following factors: irrigated or tree-planted land; non-irrigated land; and size of family. Thus the maximum of 30 hectares was given for 'dry' land and 8 hectares if irrigated. Any land which was uncultivated or not allocated was 'free' for rental to those prepared to farm it with credit granted by the nationalised banks for the purchase of materials: seed, fertilisers, and machinery. The development of agricultural co-operatives has been but a natural consequence of the continuation of agrarian reform.

Those who have received land have gained, as the annual payments of the purchase price amount to only one-sixteenth of the rent previously paid to landlords.

The poorest former tenants have obtained the most benefits, with the results of the reform clearly being seen in the old, more crowded regions, and to a much lesser degree in areas of recent settlement with a sparser population. Obviously in order to maintain strict observance of land tenure rights, the expensive operation of resettling farmers from areas of dense population in those of lesser density would be the answer. A theoretical solution, but in practice probably unworkable.

Historically, such was the character of land ownership that it was normal for the farmer, and not the owner, to be the person who actually worked the land. The system of joint cultivation was the predominant system of agricultural exploitation at that time, particularly in the unwatered or rain-fed lands. But there was no legislation protecting the farmers from being exploited by the owners who, as a class, took the lion's share of the production.

Four kinds of land-tenure were typical in Syria.

Direct Exploitation

The owner, with the help of his family or of hired labour, enjoyed full control of his own land and farmed it entirely for his own benefit. This system obtains in smallholdings, particularly in mountainous country and in areas close to towns and villages.

Cash Rental

This system was extremely limited in Syria, since it required that the farmer should earn the means both to pay the rent and to bear all the necessary agricultural expenses, which in most cases he was unable to meet without serious detriment to his standard of living. Where this form is still found, the amount of the rent differs from one region to another, according to the nature of the land and to whether water-supply comes from rain or from irrigation.

Joint Cultivation

Though this differed in detail from region to region, the basic principle was that the farmer worked the land in return for a share of the produce, the size of his share varying according to circumstances. In medium-sized holdings where for one reason or another the owner chose not to work, the farmer, at the owner's request, took full charge of the farming operations and paid the owner a quarter of the produce, in cash or in kind. This arrangement was termed the *morabaa*, or quarter system.

An even more widespread arrangement was the *moshanka*, or participation system, under which the owner provides the land, the farmer does all the work, and they split the produce between them, sharing profits and losses. Owners of large areas of non-irrigated land prefer this system because the success of agriculture in these areas depends on rainfall which can vary from year to year, so that the landowner is not paying for hired labour to do work which might have uncertain results. At the same time his risk of loss is limited, since it is the farmer who bears the cost of the seed, machinery and animals in addition to contributing his own and his family's labour. The division of profits between landowner and farmer differs according to the density of population of the area as well as the regularity of the rainfall.

This archaic system was the underlying cause of the poverty of the older generations of Syrian farmers. Under this arrangement, if the season was bad, the farmer was forced to borrow money from the landowner at a very high rate of interest simply to keep his family from starving. In this way, the landowner was in a position to tie the farmer to the land for as long as he was indebted to him, and to force the farmer to work under his dictation.

In the Gezira, where agriculture is largely dependent on modern irrigation involving the use of machinery, partnership is formed by three parties, the landowner, the owner of the irrigation machinery, and the farmer. These parties share the profits, usually in the proportions of 25 per cent, 50 per cent and 25 per cent respectively, though these rates sometimes differ.

Common Land

In parts of Syria, particularly in the regions east of Aleppo and Homs and in the Houran region in the south, there is a form of collective ownership of land. All the land in and around the village is the joint property of the inhabitants, each individual having a common right to the land. The land is classified according to its situation and productivity, and each class of land is divided into small parcels and allocated to individual villagers. Exploitation of the land is individual, even if ownership is collective, while the allocation of land among farmers is not permanent but comes up for re-allocation every three or four years.

AREAS OF AGRICULTURAL PRODUCTION

Syria may be divided into seven main agricultural regions.

1. The Houran Plain and Jebel Druze Region

This region lies in the extreme south of Syria. In ancient times the Houran Plain was famed for its fertility and the Romans regarded it as a vast granary. Because of soil exhaustion, however, it is not so fertile now. The most important grain crops grown in the region are wheat, millet and different varieties of maize.

South of the plain of Houran lies the upland area of Jebel Druze, with its covering of fertile volcanic soil which favours the cultivation of grain. Here smallholdings predominate and there is a high density of population. Towards the west, fertile cultivated land is mostly 'dry' and produces good crops of cereals, although eastwards the countryside is soilless and barren. Rainfall fluctuates widely from year to year, however, which renders harvests uncertain. The conditions are aggravated by the fact that subterranean water is scarce, even for drinking. To the north-west of the Houran Plain rise the Golan Heights, a rocky, uncultivated area suitable only for pasture, though orchards flourish where there is enough water.

2. *The Damascus Sector*

This region lies north and south of the city of Damascus. It is an area of semi-desert, the rainfall being insufficient for agriculture. However, the exploitation of the waters of the Barada River has created a large oasis called the Damascus Ghouta, an area of very fertile land supporting numerous vineyards and olive groves, the production of which has greatly increased in recent years. Dairy-farming is also widespread.

Medium-staple cotton being picked in the Hama region.

3. The Homs, Hama and Aleppo Plain

Between Homs and Aleppo there is a wide plain with an average annual rainfall ranging from about 600 mm in Homs to about 400 mm in Hama and Aleppo. This plain (apart from the Gezira region) is the most important area for agriculture in the whole of Syria. In it there are a number of agricultural expansion projects involving the use of irrigation. In the Homs and Hama region a network of irrigation channels fed by Lake Homs brings water to about 22,000 hectares. Large areas are watered by wells and by the water of the Orontes river, which is raised by water-wheels and pumps. Among the more important undertakings in this region is the Al-Ghab scheme. Completed in 1965, it increased the irrigated area by some 70,000 hectares (see map).

Swamps in the al-Ghab lowland, before drainage and reclamation of the land made this area one of the most fertile in the country.

The Homs, Hama and Aleppo Plain has become the centre for cotton cultivation in Syria. It also grows about a third of the country's production of wheat and half its output of millet. In western Aleppo province, the cultivation of vines, olives and pistachio nuts is widespread.

There is a noticeable variation in the type of soil; whereas in some areas it is poor and granular, elsewhere it is red and fertile. The productivity of the land therefore varies according to the type of soil, as well as the amount of rainfall. Large land-holdings based on the participation or *moshanka* system were predominant, and although mechanisation is appearing in the region, traditional methods are still in use on most farms.

About 12-15 million tons of oil a year are produced in Syria, and what, in view of the chronic world shortage, is a vitally important nitrogenous fertiliser plant is now in production at Homs. Phosphate factories are increasing in numbers with the injection of foreign capital for the insatiable needs of soil improvement, on which depends the production of crops to feed the ever-increasing population.

4. The Tadmor (Palmyra) Basins

These are a number of oases in the centre of Syria and the most important of them are the Gairud and the Tadmor. The Tadmor oasis was famous for its produce in Greek and Roman times. Grain and fruit cultivation is carried on in the oases, which depend on underground water. Research has revealed the presence of considerable subterranean water-resources which will make it possible to extend the cultivated area round these oases.

5. The Northern Plains

These form a productive area 50 to 100 kilometres wide, shaped like an arrow-head and surrounded by the Syrian desert. Here the soil is dark and reddish, becoming deeper and more fertile as one moves northwards and thinning out on the fringes of the desert. Large landholdings prevail. Before the agricultural reform laws these ranged between hundreds of hectares in the central plains and tens of thousands of hectares in the Gezira. There are also large tracts of state-owned land in these plains.

Agriculture is sparser here than in the regions previously described. A shortage of manpower has led to the use of up-to-date agricultural machinery, particularly in the newly-developed area of the Gezira, which now contributes 4 per cent of Syria's wheat production. Areas of good soil are intensively farmed but there are still large tracts of moderately good soil which, with a little reclamation, could be brought into cultivation.

In the Darik area, in the extreme north-east corner of the Gezira, a rich, black or dark brown soil is to be found. This area was once famous for the growing of rice but in recent years this has given way to the increased cultivation of cotton. Vines, potatoes, tomatoes and other vegetables can be successfully grown, but expansion has been held back by the remoteness of the area and the inadequacy of transport and storage facilities. A large number of rivers flow into the region and these can be utilised for irrigation, to bring added prosperity.

6. *The Euphrates and Khabur Valleys*

With their extremely fertile alluvial soil, these valleys could become very productive if supplied with sufficient irrigation water. This is the region with the highest potential for agricultural expansion in the whole of Syria. Ownerships here range from large estates with absentee landlords and cultivated by farmers on the *moshanka* or participation system, to smallholdings farmed directly by owner-occupiers. The system of agricultural rotation practised here is that usually adopted in areas of unwatered cultivation. Some lands directly situated on the river-banks are irrigated, but only for part of the year.

7. *The Coastal Plains*

These lie in the north-western part of Syria in the region of Tartus and Latakia and include the lowland area around the Ansariye Mountains. It is the rainiest part of the country, the average annual rainfall ranging between 600 and 1,000 mm. The area is suitable for the cultivation of grain, and three-quarters of Syria's tobacco crop is grown here. Other main crops are citrus fruits, vegetables and winter grain. Although there is enough manpower here, the area of land under cultivation is limited. Considerable future expansion can be achieved by using for irrigation the waters of numerous rivers which flow from the Ansariye uplift through the plains to the sea. Small landholdings, seldom exceeding 10 hectares, prevail in this area. The population is dense, particularly in the lands surrounding the hills. There is a limited use of agricultural machinery, with traditional methods still practised.

Hemp drying, for manufacture into twine and rope.

AGRICULTURAL DEVELOPMENT PROBLEMS

The Maldistribution of Land-Ownership

We have seen how the maldistribution of the ownership of land has in the past been one of the main factors hindering agricultural development in Syria.

The Agricultural Reform Law, introduced in 1958, limited private land-ownership to a maximum of 200 acres in dry areas. Any land above this figure was requisitioned and redistributed among the landless farmers, compensation being paid to the dispossessed owners. As for the land that was redistributed, the law provided that no farmer should receive more than 8 hectares of irrigated and 30 hectares of unirrigated land.

Fluctuations in Productivity

Agriculture in Syria, as we have noted, is mainly dependent upon rainfall. The area depending upon irrigation is around 14 per cent of the whole agricultural area. The Euphrates Dam should add 500,000 hectares to the irrigated area by 1990. However, little is known about the development of the Euphrates irrigation; the soils accessible are rather limited and possibly of a shallow nature which would not permit lengthy periods of irrigation. The annual rainfall varies considerably from year to year and the times of the year at which it falls are also irregular. Consequently agriculture is exposed to violent fluctuations, with periods of drought of varying severity. Very dry spells occurred in 1947, 1952, 1955, 1958 and 1960. In years of drought, production is reduced to a third or a quarter of what it is in a good year. Conversely, 1972 was a record crop year. Fluctuations of temperature and of winds also affect productivity. When all these unfavourable elements occur together, as in some years they do, great damage is done to crops, and the resulting shortage in production reflects most adversely on the national economy of the country as a whole.

Level of Productivity

In Syria the average yield per acre of grain is lower than the world average. However, the Third Five-Year Plan (1971–75) has stepped up investment in agricultural schemes to 34·8 per cent, and this has led to a large increase in the country's productive capacity. The progress achieved in agriculture in recent years is due almost entirely to the expansion of the area under cultivation and not to any great extent to a rise in rates of productivity. The low level of productivity is due to a number of causes: poor soil; the non-use of modern scientific methods in farming, particularly where quality of seed and quantity of fertilisers are concerned; and the inequitable distribution of ownership of land which in itself is the reason for the wide differences in agricultural incomes. The total cultivated area is now about 2·8 million hectares, of which around 75 per cent is devoted to cereals and dry-farmed legumes.

High Production Costs

Production costs in Syrian agriculture are high. In irrigated land, the annual

Before sundown, nomads herd their goats and sheep together after a day's foraging on pastures on the fringes of the Syrian Desert.

Part of the headworks of the Euphrates Dam in the north-east. The completion of the super hydro-electric power system will develop further Syrian agriculture, greatly increasing the importance of the Gezira as irrigation and land reclamation proceed.

cost of supplying water to 1 hectare is 15 per cent of the value of the crop it produces. Forty per cent of this land is watered by pumps, while wages for ploughing and harvesting by machinery are high, particularly in the Gezira and Euphrates regions. Facilities for storage are still inadequate, with the result that many farmers are obliged either to market their produce immediately it is harvested or to conclude contracts with merchants beforehand; so that under this system the cultivator never obtains the best competitive prices for his crops.

Above all, it would seem that one of the main factors hindering agricultural progress in Syria is the insufficiency of means of transport. Most of her rivers are not navigable, while her railways and roads have been built mainly for the purpose of linking the big towns and cities with one another. A clear example of this inadequacy is that the Gezira and Euphrates areas, which grow about half the country's total production of wheat and cotton and a third of her total output of millet, are not connected by any modern means of transport to the port of Latakia. Because of this, the cost of transport accounts for about 40 per cent of the export price of wheat and 50 per cent of that of millet. The provision of an adequate system of transport is a top priority.

AGRICULTURAL CO-OPERATIVES

The year 1970 has become known as 'the year of co-operatives', for during it more than a thousand were set up. These can be classified as follows:

Multipurpose agricultural co-operatives	978
Agricultural co-operatives for sheep-rearing	39
Agricultural co-operatives for cattle-rearing	17
Joint agricultural co-operatives for marketing	10
Agricultural co-operatives for fisheries	6
Agricultural co-operatives for sheep-fattening	5
Agricultural co-operatives for producing and pressing vegetable oils	2
Productive agricultural co-operatives (collective work)	1
Total	**1,058**

Co-operatives	No. of villages	No. of co-opera-tives	No. of members	Paid up capital £ sterling	Subscribed capital £ sterling
Agricultural	734	578	36,304	145,400	226,800
Agrarian reform	603	480	26,045	50,600	105,000
Total	1,337	1,058	62,349	196,000	331,800

There are some thirteen agricultural co-operative unions in the provinces and a general agricultural co-operative union in Damascus.

Agricultural Mechanisation

More use is made of modern mechanical means of production (such as tractors, combine harvesters and irrigation pumps) in agrarian reform co-operatives than in other individual co-operatives. In the reform co-operatives these machines are owned by the co-operatives and not by the members, whereas, by contrast, in the other co-operatives the relatively few machines are mostly owned by the members. The State, as a major participant in the agrarian reform organisation, has initiated many agricultural projects, particularly irrigation schemes in villages where there are agrarian reform co-operatives which, with previously existing projects, were taken over with the land and, on redistribution, became the property of the members.

The current policy of the State is to foster the spread of mechanisation generally in agriculture and to grant special privileges like credit facilities to co-operatives in furthering this aim.

Agricultural Production

The majority of agricultural co-operatives do not specialise in any particular crop. The area they cover amounts to no less than 335 hectares, of which nearly 120,000 are irrigated, and 215,000 unwatered; and they cover, in addition, about 44,000 hectares of forest land.

The following table shows the major crops grown in 1973 and 1974:

Area and Production of Principal Crops in Syria

	1973		1974	
	(hectares)	(metric tons)	(hectares)	(metric tons)
Wheat	1,476,000	593,000	1,537,220	1,629,896
Barley	914,000	102,000	696,952	655,480
Maize	11,600	15,400	13,500	19,220
Millet	21,600	12,500	26,300	14,203
Lentils	92,100	23,700	85,411	83,369
Olives	175,000	73,000	179,834	215,010
Cotton	200,400	404,300	205,474	386,534
Sugar-beet	7,700	152,400	6,500	138,900
Tomatoes	20,700	269,000	29,883	395,467

Source: *The Middle East and North Africa 1975–76*, Europa Publications, Twenty-second edition.

Co-operative Marketing

The marketing of agricultural crops in the co-operatives is carried out either directly or indirectly: sometimes a co-operative sells its crops direct to a purchaser or to a government organisation, as is usually the case with cotton, grain, tobacco and sugar beet; but in other cases its crop is marketed by joint co-operatives, as is usual in the case of sweet potatoes, tomatoes, grapes, groundnuts and vegetables generally. Co-operative marketing is developing year by year. The State is increasing its material assistance and exerts much effort to encourage the spread of co-operative marketing, since this is the only way to rid the producer of unfair exploitation by middlemen and to arrive at a price which is reasonable for both the producer and the consumer.

THE NATIONAL INCOME FROM AGRICULTURE

One of the major problems with which the Syrian Government has had to deal is the under-use of cultivable land. It must be conceded that habit constitutes a strong force in the practice of keeping about a quarter of this land fallow in the process of crop rotation and soil conservation. Nevertheless, the very high figure of 62 per cent of cultivable land remaining unused in 1970 can only partially be explained by lack of water—the Euphrates Dam should do much to ameliorate this problem in the future.

When horizontal expansion reached its peak in 1960, it became necessary to concentrate on vertical expansion to increase the contribution of agriculture to the national income. This was undertaken in the following ways:

by bringing irrigation to hitherto unwatered land and by developing newly-irrigated areas;

by improving the system of crop-rotation in order to reduce the percentage of unproductive land at any given time, particularly in areas of irrigated land which also had abundant rainfall;

by encouraging generally the greater use of agricultural aids such as fertilisers, hardier strains of seeds, pest-control machinery and improved animal breeding;

by employing a greater and more efficient manpower force to intensify agricultural output as a whole.

Each of these measures is effective in itself, but when all are integrated into a single driving effort, one element often serves to increase the effectiveness of another and thus of the whole.

So much for the economic aspects of government policy. Let us now look at the issue from the social point of view, and in particular at the redistribution of land and its consequences.

Before the implementation of the agrarian reform laws at the end of 1958, the inequitable distribution of agricultural land in Syria was one of the main problems facing the country. The large landowners held power by virtue of their feudal, semi-feudal and semi-capitalist rights, and a totally obsolete system of trade relations prevailed. Traditionally large estates have many shortcomings. Politically they mean domination and oppression, socially they create drastic differences in incomes, technically they tend to lead to low productivity, and economically they result in the owners grasping a large part of the surplus and squandering it in unproductive ways. Thus it became necessary to sequestrate the large landholdings and eradicate the counter-productive relationships they entailed.

We have seen how the first agrarian reform law, which was issued a few days after the unification of Syria and Egypt in 1958, fixed limits to the amount of land an individual could possess and how, following the modification to that law in 1963, these limits were changed. In 1966, Law No. 88 was issued and made the following provisions:

1. Further to limit the maximum area of land held by any one owner, and at the same time to classify regions according to their productivity. In irrigated or forest areas, ownership ranged between 10 and 50 hectares (as against 80 hectares allowed under the original law of 1958); and in unwatered areas owners could hold between 80 and 200 hectares (as compared with 300 hectares allowed by the 1958 law). An exception was made in regard to areas in the north-eastern provinces in which the rainfall is less than 300 mm.: here the maximum holding remained at 300 hectares. This legislation also provided that the maximum allowed to an owner should be increased by 8 per cent for a spouse and for each child.

2. Farmers receiving the redistributed land could each own a maximum of 8 hectares of irrigated or forest land and 30 hectares of unwatered land. Agricultural reform co-operatives were also set up for those farmers receiving redistributed land.

3. Farmers receiving redistributed land were exempted from paying rent for it, but they had to purchase it at a price assessed on the basis of one quarter of the value of the compensation payable by the State to the dispossessed owner, this price to be paid by the farmer in equal annual instalments spread over twenty years to the agricultural co-operative fund in order, firstly, to finance agricultural projects and, secondly, so that members of the co-operative might benefit from it.

4. The new legislation enabled collective farms to be set up wherever necessary. Until 1969 nearly 800,000 hectares had been redistributed among 52,500 farmers who, with their families, brought the number of individuals benefiting from the reforms up to 300,000, with an average of 15 hectares per family.

Of the 750,000 hectares requisitioned but remaining undistributed, some 350,000 hectares were allotted to State farms, while the balance, mostly unirrigated land, will not be distributed until a suitable form of agricultural use can be devised for it.

Thus the agrarian reforms in Syria have mainly concentrated on the question of bringing about a more equitable distribution of land. Let us now consider the importance of agriculture in the lives of the people.

As we have already seen, in 1970 the population of Syria was estimated at 6,924,000, with a rapid urban growth (57 per cent) and 29 per cent forming the rural population, the great majority of whom live by agriculture.

Thus the population of Syria is growing at the high rate of 3·3 per cent annually, the rate of growth being slightly higher in urban districts, while the proportion of the rural population is slowly declining. Since the horizontal expansion of agriculture has reached its limit, the high rate of population growth is leading to an increasing pressure on cultivated land, as the following table shows:

Growth of Population and Development of Cultivated Land in Syria, 1946–1970

Year	Cultivated land (hectares)	Population	Cultivated land per capita (hectares)
1946	1,800,000	3,006,000	0·60
1956	3,100,000	4,025,000	0·77
1960	3,500,000	4,841,000	0·73
1966	3,127,000	5,761,000	0·54
1970	3,100,000	6,924,000	0·45

Source: Syria, Ministry of Planning, *Statistical Abstract*, Damascus.

Furthermore, this rapid growth of population will aggravate the problem of unemployment in all its forms; it will increase the demand for foodstuffs and other commodities, and it will create problems of housing and other social problems which already constitute obstacles in the way of progress, including the mechanisation of agriculture, research, and so on. Efforts are being made in various fields to overcome the effects of this adverse trend; for example, by bringing more fallow land under cultivation each year with the aim of reducing by half the present three million hectares of fallow to 1·5 million by 1980; by expanding the area of irrigated land; by encouraging the breeding of more animals and poultry; and by industrialising the rural areas and improving buildings and communications. Industrialisation in itself constitutes one of the most important and effective solutions to the problems of population growth.

REFERENCES

The Middle East and North Africa 1973–74, 20th edition, Europa Publications
Marion Clawson, Hans H. Landsberg, Lyle T. Alexander, *The Agricultural Potential of the Middle East. Economic and Political Problems and Prospects*, American Elsevier, New York 1971
J. I. Clarke and W. B. Fisher, *Populations of the Middle East and North Africa : A Geographical Approach*, University of London Press, 1972
Michael Adams (Ed.), *The Middle East : A Handbook*, Anthony Blond, 1971

Palestine

In the Preface to this book, I explained why no chapter is devoted to Israel. I should now like to go more fully into what all Arabs see as a burning political injustice.

HISTORICAL SUMMARY

The name Palestine is derived from that of the Philistines, who occupied the southern coastal region of the Levant in the twelfth century BC. Up to that time, Palestine had been inhabited by the semi-nomadic Semites and the Canaanites, who had settled in the coastal plains as early as the twentieth century BC. The first Jewish presence in Palestine began at the time of Abraham, the leader of a nomadic Semite tribe who migrated from Mesopotamia and settled in the area around the present-day Hebron. Here they found an established people composed of Canaanites, Jebusites and Philistines from whom they were unable to wrest control of the Mediterranean coast. They were, however, able to incorporate inland areas into a Kingdom of Israel, which was subsequently split into two when the southern part became the Kingdom of Judah. During the next 250 years, the northern kingdom was dominated and then overwhelmed by the Assyrians and in 721 BC Israel became politically extinct. The Kingdom of Judah was also brought to an end when Jerusalem fell to the Babylonians in 538 BC.

Between 166 and 63 BC the Israelites were, however, able to form the Maccabean kingdom in part of Palestine and to enjoy political independence once again. During this period the Temple and the walls of Jerusalem were rebuilt, as had been predicted by the Prophets during the Babylonian Exile. The ancient kingdom of the Jews finally fell to the might of the Roman Empire in 63 BC. From then on, attempts to gain independence continued until AD 135, culminating in a tremendous revolt which the Romans suppressed, bringing their religion into the Temple of Jerusalem. The waning of political influence continued until the beginning of the Zionist movement. On the other hand, the Palestinian Arab emerged as the representative of an indigenous people who had been continuously present in Palestine since the twentieth century BC, some eight centuries before the first Jewish influx.

In AD 1882 there arose the vision of yet another return of Jews from exile, this time from their prolonged and voluntary exile. This was the dream of the Zionists: those who advocated the colonising of Palestine by modern Jewry.

The fifteen years 1882 to 1897 saw the beginning of Jewish colonisation; but the early sporadic attempts to implant a Zionist settlement-community in Palestine proved such a failure that a serious reappraisal of strategy was necessary. This was finally accomplished by Theodor Herzl, the founder of Zionism as a political movement in 1897. At the conclusion of the first Zionist Congress at Basle in 1889, Herzl confidently wrote in his diary, 'Were I to sum up the Basle Congress in one word—which I shall guard against pronouncing publicly—it would be this: at Basle I founded the Jewish State. If I were to say this out loud today, I would be answered by universal laughter. Perhaps in five years, and certainly in fifty, everyone will know it.' In fifty years everyone did know it. Herzl's Jewish State had been founded in spite of the intrinsic rights of the Palestinian Arabs and in spite of the unyielding resistance they continue to offer against the usurpers of their land.

By the outbreak of the First World War, however, the Zionist colonisation of Palestine had met with only partial success after some thirty years of effort on the part of the Zionists, who represented an infinitesimal minority (about 1 per cent) of the entire Jewish population throughout the world. Their activities had aroused the outright opposition of other Jews, who sought to solve the so-called Jewish problem by assimilation in Western Europe and the United States, and not by 'self-segregation' in Palestine. The indigenous Jews of Palestine constituted at this time just 8 per cent of the total population of the country and possessed no more than $2\frac{1}{2}$ per cent of the cultivable land. For all its financial support and early confidence, Zionism had yet to obtain any form of political endorsement from the Ottoman authorities then governing Palestine or, indeed, from even one of the major European powers.

The outcome of the First World War finally set the stage for an alliance between the British Government of the day and the Zionist Movement which, during the subsequent thirty years, succeeded in accomplishing the objectives laid down by Herzl and paved the way for the dispossession and expulsion of the Arab people of Palestine and the creation of the State of Israel in 1948. Unlike the traditional European colonisation elsewhere, the Zionist colonisation of Palestine was essentially incompatible with the continued existence of the native population in the coveted country.

In 1915 Britain had promised Sheikh Hussein of Mecca that, in return for the help of the Arab armies in the campaign against Germany's major ally Turkey, all Palestine, Iraq, Syria and the Arabian Peninsula would be free and independent once the Turks had been defeated. No sooner had this pledge been given and the Arab armies had mobilised in response, than Britain and France combined to sign the infamous Sykes–Picot agreement of 1916 and agreed to share out Syria, Iraq and Palestine between them as the spoils of war. In addition, in 1917 the British Secretary of State for Foreign Affairs, Arthur James Balfour, issued on behalf of Britain's wartime Government the Declaration that was to bear his name: 'His Majesty's Government view with favour the establishment in Palestine of a national home for the Jewish people and will use their best endeavours to facilitate the achievement of this object, it being clearly understood that nothing shall be done to prejudice the civil and religious rights of the existing non-Jewish communities in Palestine.'

Having acquired and then relinquished these spoils in so short a time, Britain then began to allow the creation of a situation whereby the Jewish immigrants could come to control Palestine for themselves. A census of the population of mandated Palestine—occupying a total land area of 27,190 square kilometres—gave the following figures in 1919:

Moslems	574,000	82%
Christians	70,000	10%
Jews	56,000	8%
Total	700,000	100%

By 1922 the population had increased to 757,000, with the Jewish population having grown to 83,794, or 11 per cent. In 1931 a new census revealed that the population of Palestine had surpassed the one million mark, with 174,610 Jewish inhabitants, or 17 per cent of the total.

In 1948 the census taken by the Palestine Government was as follows:

Moslems	1,972,560	69%
Christians	145,060	5%
Jews	608,230	21%
Others	151,490	5%
Total	2,877,340	100%

Thus the proportion of Jewish inhabitants had risen from 11 per cent in 1922 to about 17 per cent in 1931, and to 21 per cent in 1948. The growth of the Jewish element is all the more startling when we consider the fact that the birth-rate of the Palestinian Arab community was nearly 50 per cent higher than that of the Palestinian Jews (3·2 per cent and 2·2 per cent respectively). Hence it can be stated with conviction that it was the large-scale immigration into this semi-fertile country that accounted for the rapid growth of the Jewish community.

At the time of the Balfour Declaration, the Jewish community owned just 2 per cent of the total land area. During the ensuing thirty years, the purchase of additional property had brought the Jewish holding on the termination of the mandate in May 1948 to 5·67 per cent of the total land area of the country. This figure was later verified by the Palestine Government, which ascribed more than 15 per cent of the cultivable land of Palestine to the Jewish inhabitants.

Arab resistance to the sale of land to the Jews persisted during the mandate period, and the additional area of 85,000 hectares acquired between 1918 and 1948 was purchased in the main from absentee foreign landowners. The actual land area relinquished by the Palestinian Arabs during the period amounted to only 40,000 hectares in spite of the high prices offered to them and the legislative measures that were specially designed before 1939 to facilitate the transfer of land, as Balfour had implied.

The British Government unilaterally decided to bring the mandate to an end in May 1948. The Jewish immigrants, under the guidance of Zionist factions, had already developed the para-military formations which were to create a state of terror amongst the Arab population and to cause a mass emigration, not only from the territories assigned to the Jews under the terms of the United Nations resolution of 1948, but from cities, towns and villages such as Jaffa and Acre which had been assigned to the Arabs.

On 15 May 1948, the State of Israel was founded. The concept of a national home for the Jewish people had been converted into a State of Israel. Against all the principles laid down under the United Nations Charter, a state had been created on religious grounds, and a minority had been granted complete control of the Arab majority. Yet Israel was admitted to the United Nations, albeit under conditions, the chief of which was that the exiled Arab refugees be permitted to return to their homes. The number of Arabs who had left their homes by 14 May 1948 was in the neighbourhood of 300,000. By the time the final Armistice Agreement was signed on 12 July 1949, after the Arab States had rallied to defend the rights of the Palestinian Arabs, the total number of refugees who had been forcibly expelled from their homes was officially estimated at 726,000.

According to the 1966–67 report of the United Nations Relief and Works Agency (UNRWA), the number of refugees registered with the Agency (on 31 May 1967) had risen through natural population increase to some 1,346,086, of whom 861,118 were in receipt of minimal rations. These figures, however, did not include the Palestinians who had lost their livelihood but not their homes, and therefore did not qualify for relief under UNRWA regulations. Moreover, the figures did not take into account those whose family circumstances permitted them to re-establish themselves without assistance in neighbouring Arab countries or elsewhere in the world. As a consequence of the war of June 1967, the Palestine refugee population was raised to about $2\frac{1}{2}$ million people.

But the nationalist feeling of the Palestinian Arabs had begun to express itself some two years before the June War. In January 1965, a new phenomenon appeared in the shape of a young and vibrant generation emerging from the tents in the refugee camps. The child born as a refugee in 1948 had reached the age of manhood and was ready to fight as a soldier to re-establish his national identity and pride. Seeing no future in the resolutions of the United Nations and unable to obtain a reply from the world as to why such a grave injustice had not been rectified, the young Palestinian decided to take the matter into his own hands. Thus the Palestine freedom fighter emerged to restore the State of Palestine on democratic principles and to permit the original Arab inhabitants to return to their native land in accordance with the expressed will of the United Nations.

In view of the violent events which have taken place in this part of the Arab lands since 1948 it is difficult to make a descriptive study of this area under present conditions. Since June 1967 new parts of Palestine and Jordan have fallen under Israeli occupation. However, these events are probably of an historically transient nature, as the War of October 1973 indicates. Indeed, such political and military upheavals have often affected this region in the course of its long history, but the situation has always been restored to normal after the termination of recurrent and relatively short-lived conquests.

THE LAND

We must refrain from describing the occupied territories in detail, as they are still affected by these events. However, we give a general picture of the land of Palestine, which contains three topographically distinct areas. The westernmost of these, the Valley of the River Jordan, will be dealt with in the next chapter, since the waters of the River Jordan form a special topic, relevant particularly to the Kingdom of Jordan. The other two areas are the Coastal Plain and the Western Plateau.

The Coastal Plain

This plain extends from Haifa southwards to the borders of Egypt at Gaza, where it joins the greater plain leading to the Sinai Desert. Its average elevation does not exceed 200 metres above sea-level and it varies in width from a few hundred metres near Haifa to more than 25 kilometres at its southern extremity. It is divided by the spur of Mount Carmel near the sea into the Acre Plain to the north and the more extensive Plain of Sharon to the south. Several wadis (which usually flow only during the winter rainy season) dissect the Sharon Plain, including the rivers al-Makba, al-Ouja and Rubin. Indeed, the al-Ouja, which flows throughout the year, flows from the neighbourhood of Prah Tekvah (Petah Tiqwa), enters the Mediterranean to the north of Tel Aviv, but only averages $8\frac{1}{2}$ cubic metres per second, which limits its utility for irrigation.

The coastal plain is mostly fertile, the Bay of Acre, formerly marshland which has had to be artificially drained and reclaimed, being now very productive. Rainfall varies from one part of Palestine to another. For instance, Galilee has an average of 1,000 mm; the amount decreases southwards until as little as 250 mm or even less falls in the Negev and the Plain of Gaza. On the whole, farms in Palestine make use of rainfall during the winter and irrigation during the summer months. Subterranean water is also available at various depths ranging from a few metres to more than 150 metres.

The western part of the plain, between Haifa and Jaffa, is suitable for the cultivation of citrus fruit, whereas the northern, eastern and southern parts are more suited to cereals (wheat, barley and maize), as well as to mixed cultivation.

Palestine has diversified her crop production so that 34 per cent is devoted to crops which are not under the heading of field crops; mainly orchards, citrus and such like (two-thirds); vegetables and potatoes (one quarter). These crops give a much higher return per hectare than, say, wheat, to which some 18 per cent of land is given over (the largest area for one type of crop).

The salt content of the soils is at present levels not harmful to crop roots; the bulk of the water is under a centralised system of control and is piped to areas, which obviously makes for more efficiency of use. Salt problems are in general confined to smaller areas such as the Jordan Valley.

In view of its fertility, this plain was the most densely populated area of Palestine. Before the Israeli occupation, its population was estimated to be 700,000—including 380,000 Jews.

226

The Western Plateau

This fairly high plateau, ranging in width between 40 and 70 kilometres, is often called the 'Palestine Mountains'. It is formed mainly of limestone, with outcrops here and there of basalt.

To the north of the Bay of Acre the plateau is broken by a fault which forms a fertile basin, the western part of which is known as the Merj Ibn-Amer. This is a triangular basin connected with the coastal plain of Acre by a narrow valley in which runs the River Qishon, on which a reservoir has been built for irrigation and other purposes. To the east of Afula this strip narrows down and extends eastwards to the Jordan Valley near Bezeq. This eastern part is known as the Vale of Jezril.

The Vale of Jezril and the Merj Ibn-Amer in fact divide the Western Plateau into the al-Jalil Plateau to the north and the Samarian Plateau to the south. To the extreme south, however, it is known as the Judaean Plateau, which ends at the Negev.

Let us consider these sub-areas of the Western Plateau in more detail:

Al-Jalil

This plateau comprises the whole of northern Palestine with the exception of the narrow plain of Acre and the Ghor of Jordan. It rises towards the extreme north where it joins the Lebanese Mountains. This area enjoys the highest annual rainfall, which ranges between 500 and 700 mm and is suitable for most types of farming. The cultivable area is estimated at 51 per cent and for this reason the Jewish immigrants tried during the British mandate to concentrate their settlements in this region. There are a few oak trees, though eucalyptus are more common: coniferous trees have taken precedence over all the others in acclimatisation to the Mediterranean type of conditions and soils.

The Vale of Jezril and Merj Ibn-Amer

These two areas constitute the most fertile region in the whole of Palestine. The Vale is made up of an area of some 16,000 hectares of which 98 per cent is cultivated, whereas Merj Ibn-Amer has an area of 88,000 hectares of which 86 per cent is suitable for agriculture. The heavy alluvial soil is formed of silt carried by the floods and streams which run down from the surrounding plateaux. The land is flat and the rainfall high and constant, so that the area is well suited to modern mechanised agriculture. The major crops include cereals, fodder and vegetables. Pear and plum trees are grown rather than citrus, whereas vines and banana trees flourish in the eastern part of the Vale.

Samaria

The Samarian upland plateau lies to the south of the plain of Jenin, being mainly limestone and some 900 metres in altitude. The plateau has been eroded to form fertile valleys with an annual rainfall between 500 and 600 mm, which enhances the agricultural potentialities of this region.

227

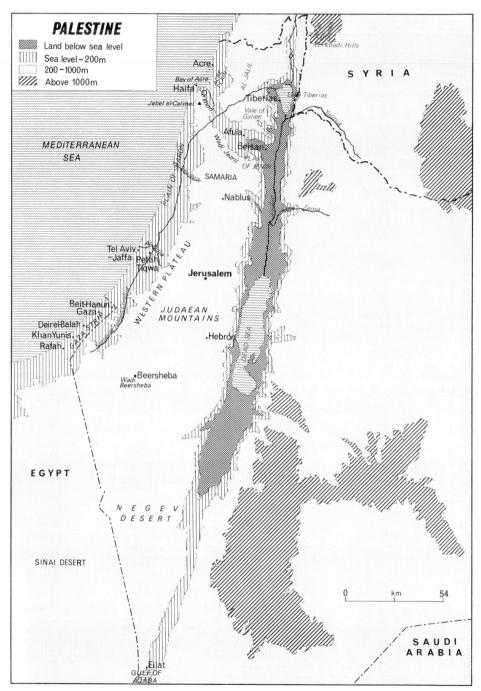

PALESTINE

- Land below sea level
- Sea level – 200m
- 200 – 1000m
- Above 1000m

SYRIA

El-Khadi Hills

Acre

ACRE PLAIN

AL JAUL

Bay of Acre
Haifa
Qishon
Tiberias
Lake Tiberias

Jebel el-Carmel ▲
Vale of Galilee

Yarmuk

MEDITERRANEAN SEA

Afula
Wadi Jezril
Beisan

PLAIN OF SHARON

OF JENIN

SAMARIA

Nahr Zerqa

•Nablus

Tel Aviv
·Jaffa   Petah
Tiqwa

WESTERN PLATEAU

Jerusalem

Beit-Hanun
Gaza·

JUDAEAN
MOUNTAINS

Deir el-Balah
KhanYunis·
Rafah·

GAZA STRIP

·Hebron

DEAD SEA

·Beersheba
Wadi
Beersheba

EGYPT

N E G E V
D E S E R T

SINAI DESERT

0   km   54

SAUDI
ARABIA

·Eilat
GULF OF
AQABA

228

The Judaean Plateau

To the south of Samaria lies Jerusalem, and el-Khalil (Hebron) on the Judaean Mountains. The western side of this plateau declines gently towards the coastal plain, whereas the eastern side falls dramatically and abruptly into the Ghor of Jordan and the Dead Sea. The plateau is deeply dissected by numerous small streams which run down the sides of the valleys after heavy rainfall in winter. The limestone terrain and surface faults allow the water to seep through and to reappear in the form of springs, which are the dominant factor in deciding the demographic settlement of the population. The total area of Samaria and Judaea together is about 7,000 square kilometres, of which 46 per cent is suitable for agricultural production. The main crops are cereals, olives, figs and grapes. As in many areas of Palestine, extensive reafforestation has taken place, with the conifer doing particularly well.

The Negev

The Negev Desert constitutes almost half the total area of Palestine and is made up of steppe and semi-desert. It measures about 12,500 square kilometres and is separated from the Judaean Plateau by the Wadi Beersheba. It forms an extensive triangle with its apex on the Gulf of Aqaba. The soil is formed of a mixture of sand and clay which makes great parts of this area suitable for agriculture where irrigation is being carried out. Rainfall, however, which ranges between 150 and 200 mm per annum, is inadequate for extensive cultivation without irrigation and the use of underground water resources. Water is also piped into the Negev from the River Jordan, via the Lake Tiberias–Negev conduit.

The Plain of Gaza

The Gaza Plain, or 'Strip' as it is better known, 45 kilometres long and varying in width between 5 and 8 kilometres, is a coastal strip on the south-eastern corner of the Mediterranean, extending from Rafah in the south to Beit Hanum in the north. Its total area is 289 square kilometres, or about 75,250 hectares.

The area contains the towns of Rafah, Khan Yunis, Deir el-Balah and Gaza, which is the capital. It also contains a number of refugee camps, as has already been mentioned. The occupants of these camps are all Palestinian refugees who were driven out of their homes in 1948, leaving their property, farms and land behind in the occupied territory.

In 1949 in a bid to help the refugees, agriculture was intensified, the total cultivated area in the Strip being then some 25,000 hectares, which was increased to 33,000 hectares by 1960. A marked expansion in the cultivation of citrus and other fruit was achieved, with a corresponding expansion in the cultivation of vegetables, cereals and tobacco. The Egyptian Administration constantly encouraged agricultural expansion and the drilling of wells for irrigation. In the area of Rafah alone, the Administration increased the cultivated area by 9,000 hectares. Plants for packing citrus fruit for export were built and large areas were afforested as a means of providing work for the refugees and of helping to ameliorate the harsh desert environment and difficult living conditions.

The Hashemite Kingdom of Jordan

Jordan, with its capital Amman and a population estimated in 1972 at 2,497,000, is in the main made up of desert with an average annual rainfall of less than 100 mm. This, in addition to the very hot summers, means that vegetation cannot grow (except at the bottoms of wadis). Therefore there is no grazing even where grey desert soils prevail throughout 50 per cent of the country. The common denominator of soils throughout Jordan is their high content of calcium carbonate.

The Jordan Valley

The Valley of the River Jordan is known as el-Ghor. The source of the river is situated in the Jebel esh-Sheikh, the part of the Jordan basin which lies within the territory of Palestine measuring about 960 square kilometres including the basin of the drained Lake Hula which is now farmland and nature reserve. From Lake Hula the Jordan flows to Lake Tiberias (the Sea of Galilee). And it is here, or to be exact just below, that the Yarmuk river, a major tributary, joins the Jordan and together they meander to the Dead Sea, 392 metres below sea-level. About 70 per cent of this area is suitable for agriculture and is under the cultivation of citrus fruit, apples, wheat, barley and animal fodder.

The Jordan Valley is the main zone with a high degree of intensive agriculture under irrigation. Estimates vary about the total capacity of suitable irrigable land—in the region of 45,000 hectares, most of which (70 per cent) is located on the eastern side of the river (see map). The main problem is that about one third of the soils of the Ghor terrace are saline and need 'leaching' treatment and drainage before agriculture can be properly and efficiently practised; once the salt is dispersed, then the soil is permanently fertile as long as a good regimen of irrigation is maintained.

Though the soils of the Zor—the lower terrace of the river—have a higher yield potential than those of the Ghor, these are subject to flooding annually during the winter months; this problem will, to a great extent, be eliminated once the Yarmuk dams are completed (see p. 208). Nevertheless there are still free-flowing wadis which enter the Jordan both to the east and the west, so that flooding to a certain extent can be a hazard.

The Jordanian Plateau

To the east of the Jordan river lies a plateau of medium height which nowhere rises more than 1,500 metres above sea-level. Nevertheless this upraised edge of the Arabian Plateau has some imposing highlands of 800 metres and, fortunately

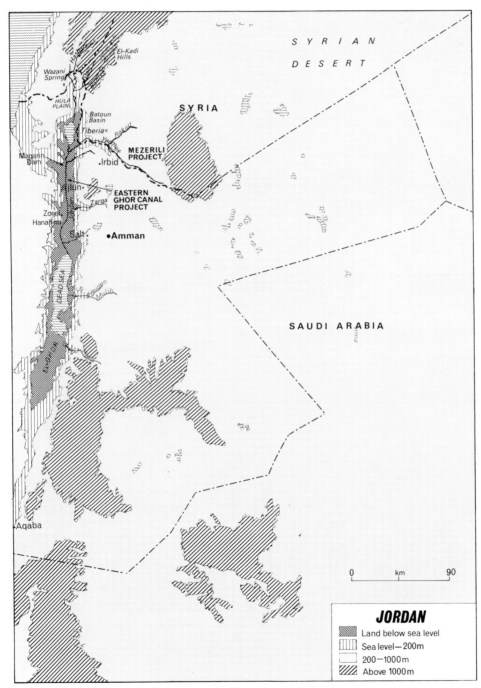

S Y R I A N

D E S E R T

El-Kadi
Hills

Wazani
Spring

HULA
PLAIN

SYRIA

Batoun
Basin

Tiberias

Maqarin
Dam

MEZERILI
PROJECT

.Irbid

Ajlun

EASTERN
GHOR CANAL
PROJECT

Zarqa

Zoret
Hanalima

Salt.

•Amman

DEAD SEA

Maiden Mulih

SAUDI ARABIA

Aqaba

0   km   90

JORDAN

	Land below sea level
	Sea level—200m
	200—1000m
	Above 1000m

for Jordan, is dissected by quite a number of streams which, as can be seen from the map on p. 239, flow westwards into the River Jordan. The plateau declines eastwards until it disappears into the Syrian Desert. The central region of this plateau is covered with a reasonably fertile soil which would be suitable for cultivation if irrigation water were to be found. However, the annual rainfall, which ranges between 200 and 400 mm, is insufficient for extensive agriculture.

The Eastern Desert

The Eastern Desert forms part of the Great Syrian Desert and makes up about 79 per cent of the total area of Jordan. It lies in the rain shadow of the highlands, and supports pastoral nomads only where artesian water can be found, as the annual rainfall is below 200 mm, which is the minimum required for cultivation. The deserts both to the south and the east support some 50,000 nomads with their sheep and goats, while here and there pockets of cultivation are to be found round oases.

LAND TENURE

There are scarcely any large individual land-ownerships in Jordan. When such ownerships are found, they are usually owned collectively by Bedouin tribes. In general, the pattern is that of small ownerships, as shown in the following table:

Number of Agricultural Holdings by Size

Size Group in Dunums	Number of Holdings	Percentages (rounded off)
Less than 10	34,039	36·4
10–19	10,193	10·9
20–29	9,363	10·0
30–39	7,621	8·1
40–49	5,393	5·8
50–99	14,221	15·2
100–199	8,003	8·6
200–499	3,747	4·6
500–999	688	0·7
1,000–1,999	198	0·2
2,000–4,999	60	0·1
5,000–9,999	16	—
10,000 and over	2	—
	93,544	100·0

Source: Jordan, Department of Statistics, *Report on Agricultural Census, 1965* (June 1967) and *Report on the Results of the Agricultural Sample Survey, 1966.* (These are the most recent reports available.)

Water melons.

The fruit of the cactus is harvested and sold as a culinary delicacy.

Results from land reclamation and the development of irrigation can be seen in the growth both of crops (using special strains of seed and inorganic fertilizers) and of citrus trees in Ghor Safi. A sprinkler system is in use for the economical application of water.

The River Yarmuk. Its waters irrigate some 12,000 hectares of arable land in the north-eastern part of the Ghor.

The River Jordan follows a meandering course through characteristic steep escarpments, with the present flood plain, El Zor, approximately 55 m below the main surface of the Ghor. The River Jordan and its main tributary the River Yarmuk account for the greater part of the water resources available for irrigation in the Eastern Ghors.

In 1960 the Government promulgated the Eastern Ghor Canal Law, aimed at helping the landless farmers, which was equivalent to the land-reform laws introduced elsewhere. This law fixed the area of the agricultural unit at between 7 and 123 hectares, providing that no family would be allowed to buy or lease more than one unit. The law also prohibited the fragmentation of agricultural units. Thus the percentage of farms with fewer than 4 hectares, which includes all types of land, is estimated at 65 per cent, these farms being worked by their owners; an additional 20 per cent (according to the official census) were farms which were part owned and part rented; and a small percentage (12 per cent) of the farmers rented their farms.

Land Use in Jordan (East Bank Only)

Forestry ('ooo dunums)		
	1973	1974
Area newly planted	12·2	25·0
Timber production (cubic metres)	603·0	568·0

Livestock ('ooo heads)		
	1973	1974
Camels	18·2	13·7
Cattle	46·3	42·2
Sheep and goats	1,351·6	1,190·2

Principal crops (metric tons)		
	1973	1974
Barley	531·7	648·7
Maize	2·0	2·6
Sesame	6·9	7·6
Wheat	2,441·8	2,462·0

Source: Jordan, Department of Statistics, *Report on Agricultural Census, 1965* (June 1967) and *Report on the Results of the Agricultural Sample Survey, 1966*. (These are the most recent reports available.)

AGRICULTURAL EXPANSION

Grazing and primitive agriculture were the mainstay of economic life in Jordan when it was created as a political entity under the name of the Emirate of Trans-jordan. The two decades following the creation of the State gave rise to a growing interest in expanding agriculture in view of improved means of communication and the growing need for agricultural products. Agricultural expansion was boosted during the Second World War and continued to grow even after the catastrophic events of 1948; but although a 15 per cent increase was achieved, the type of rain-fed cultivation which is prevalent in Jordan has its inevitable limits. Ground-water supplies can be utilised where rainfall accumulates in lower strata (permeable) and if used with control, these supplies are recharged by natural means. Water can be obtained by pumping.

Agricultural development in Jordan is thus bound to be concerned mainly with devising projects for the irrigation of larger areas in the valley of the Jordan river, the most important of which is the East Ghor irrigation project. When the two storage dams on the Yarmuk river are completed (see p. 208 in the chapter on Syria), Jordan should obtain about 550 million cubic metres of water, which represent about 72 per cent of her requirements for irrigation. More details of the problems involved in the exploitation of the Jordan river are given on pages 240 to 246 of this chapter.

Modernisation and progressive farming methods are also of great importance in the development of Jordan's agriculture. Re-seeding of grasslands for controlled grazing is one example, freeing 90,000 hectares of the arid lands previously yielding crops (fallow land has problems not only of weeds but also of moisture control, erosion, and so on). Abandoning the fallow-crop system where irrigation is not available is another, as is following an annual harvesting system for oats, vetch, beans, peas and so on, with contour ploughing and the added use of fertilisers such as nitrogen. Planting the short-straw variety of wheat, which is hardier under arid conditions, is yet another of the changes in the pattern of agriculture which are coming about in Jordan today.

The Importance of Agriculture in Jordan

Jordan has an area of approximately 94,500 square kilometres; statistics vary about the amount of cultivated land, which comprises about 10,695 square kilo-metres. Land for cultivation is highly variable, as this aspect is closely linked with rainfall, which is too unpredictable and fluctuates a great deal from year to year.

With the exception of Jordan and Palestine, agriculture is the major concern of the countries of the Middle East: the social structure is built upon it. Only in Palestine is the rural population smaller than the urban. Since 1967, however, a curious population situation has developed which could probably be described as unique, in that the numbers of refugees and displaced persons are forming the majority of the population.

The registered number of the refugee population is now about 1,532,000, according to the United Nations Relief and Works Agency for Palestine Refugees in the Near East. Many live in special camps such as those in the Gaza Strip or in

East Jordan, thousands having fled from the West Bank in 1967, causing great demographic problems for those on the East Bank.

THE WATERS OF THE JORDAN

For geographical and historical reasons an account of the division and exploitation of the precious waters of this river most suitably falls within this chapter, because the greater part of the Jordan Valley or basin lies within the territory of Palestine and the Kingdom of Jordan, although both Lebanon and Syria are vitally concerned in it too. Before the construction of the East Ghor Canal and dams and tunnels on the tributaries to harness the waters of this important river for the use of the countries involved in its exploitation, many projects were considered, later to be rejected for one reason or another; eventually a proposal or set of proposals came about which were more or less acceptable to all four countries. Where such a precious commodity as water is concerned, it is inevitable that some dissatisfaction should arise between the parties involved. Let us first examine the geography of the region.

The Upper River Jordan is formed by the confluence of four tributary streams: namely the Bareighit, the Hasbani, the Dan and the Baniyas rivers, which combine above Lake Hula. The Jordan thus draws its headwaters from streams rising in Lebanon, Syria, Jordan and the occupied territory. These rivers arise from the foot of the el-Sheikh Mountain (Mount Hermon) and of them the most important is the Dan, which rises in the hills of el-Kadi in the occupied territories. The longest is the Hasbani, the major part of which lies in Lebanon; and the shortest is the Baniyas, which rises in Syria.

Water flowing in a lined irrigation canal feeds a storage tank for the economical use of water for citrus cultivation.

These rivers meet in the occupied territories in the northern part of the Hula Plain to become the Jordan, which then flows into Lake Tiberias after being fed by various springs and streams coming down from the eastern heights in Syria and the western uplands now in the occupied territory.

The River Jordan leaves the southern shore of Lake Tiberias and runs south for a distance of 103 kilometres before it empties into the Dead Sea. In its course from Lake Tiberias to the Dead Sea the Jordan is fed by two main tributaries: the Yarmuk, most of which lies in Syria, and the Zarqa, which originates in Jordan.

Since the occurrence of the Palestine tragedy, Arab public opinion has paid much attention to the importance of the waters of the River Jordan and its tributaries. High-level discussions have taken place both in Arab capitals and in international circles on the need to exploit these waters and to protect the rights of the Arab peoples to derive from them whatever benefits there may be in the form of irrigation, industry, and hydro-electric power. This interest and concern were intensified when the Israelis violated the sovereignty and rights of the occupied territories by embarking upon a project to pump large quantities of Jordan water from Lake Tiberias to irrigate the Negev Desert. This project was devised as part of the general Israeli aggressive policy towards the Arabs.

By measuring the discharge of the waters from the River Jordan and its tributaries at certain points and by calculating the evaporation from its surface in the different sectors, especially Lake Tiberias, we obtain the following figures: the Hasbani and Bareighit rivers, which rise in Lebanon, give it 157 million cubic metres annually; the Baniyas, which rises in Syria, adds a further like quantity of 157 million cubic metres; and the Dan, which rises in the occupied territories at the foot of the el-Kadi hill, supplies 258 million cubic metres. (It was for this reason that the Israelis fought so furiously on this hill during the 1948 War, for at its foot there lies a spring from which flows this vast amount of precious water.) The springs and streams from the eastern heights provide 159 million cubic metres, and those from the occupied lands on the west a further 339 million, and the largest of all, the Yarmuk, which rises in Syria, produces 475 million cubic metres, 424 million from Syrian territory and 51 million from that part of the Yarmuk's course that lies in Jordan. Finally, the Jordanian springs and streams which flow into the river from both east and west supply about 335 million cubic metres.

These sources make up the whole of the water which flows into and out of the River Jordan annually, and amounts to an estimated 1,880 million cubic metres per annum. After deducting an estimated 630 million cubic metres for loss by evaporation, the remaining 1,250 million cubic metres flow into the Dead Sea.

Since these estimates are based on past figures, the Arab countries concerned are taking the necessary measures to obtain more precise and up-to-date figures, especially in view of the draining of the Hula swamp and the setting up of projects for the exploitation of the water along the course of the river and its tributaries. The figures quoted above show that 68 per cent of the river water, i.e. 1,283 million cubic metres, comes from Lebanon, Syria, and Jordan, of which 740 million or some 39 per cent comes from Syria alone; the remaining 32 per cent comes from the occupied territories.

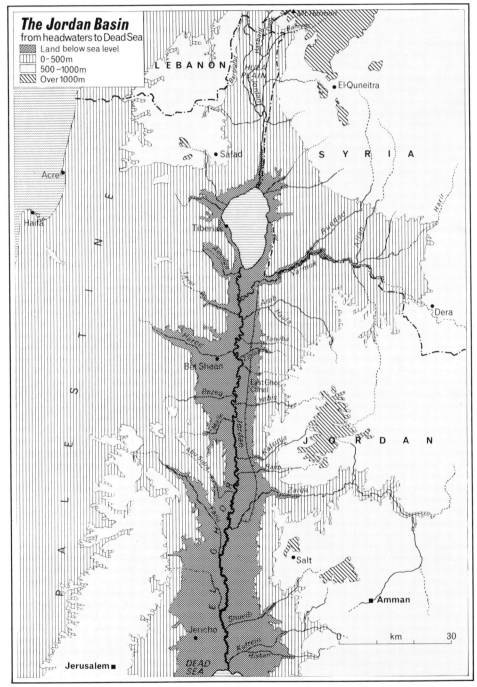

The Jordan Basin
from headwaters to Dead Sea

Land below sea level
0 - 500m
500 - 1000m
Over 1000m

LEBANON

Mt Hermon

Banias

Hasbani

Bereighit

Liddani

HULA

SWAMP

El-Quneitra

S Y R I A

Safad

Acre

Haifa

Tiberias

Ruqqad

Allan

Harir

Tavor

Yarmuk

Arab

Hadid

Dera

Jalud

Bet Shean

Haniba

East Ghor

Canal

Yabis

J O R D A N

Bizeq

Kufranja

Abu Sidra

Rajib

Zarqa

P A L E S T I N E

Salt

Amman

Jericho

0   km   30

Jerusalem

DEAD
SEA

239

THE DEVELOPMENT OF THE RIVER JORDAN

The British promise of 1917 to establish a national home for the Jews marked the beginning of a number of large-scale studies carried out by Zionist societies, Palestinian circles and other specialised bodies. These studies led to many projects being drawn up with the aim of either fostering Jewish immigration on the one hand or helping in the rehabilitation of the displaced Arabs on the other. The Zionist projects were based on the idea of accumulating all the resources in the Jordan and Yarmuk basins with the eventual purpose of controlling them so that they could be used for irrigation and electricity generation. As a reaction to the Zionists' schemes, the Arabs produced plans of their own. It is important that we should summarise here some of the background to the plans put forward by the interested parties for utilising the waters of the Jordan.

The Lord Milk Project

The American engineer, Walter Clay Lord Milk, visited Palestine as far back as 1938 on behalf of the U.S. Secretary of Agriculture to study the economic possibilities of Palestine. In his report he proposed that:

a. by using water from the Baniyas, Dan, Hasbani, Yarmuk and Zarqa rivers, canals should irrigate the plains of Beni Amer, the Bissan pastures, the Vale of Galilee, and the whole of the lowland or Ghor area;

b. the Litani river in Lebanon should be diverted across the northern region of Palestine to form an artificial lake in what is now occupied territory; thence the water would be pumped to irrigation schemes in the Negev Desert;

c. electricity should be generated by hydro-electric power, using the head of water resulting from the difference in level between the upper reaches and tributaries of the Jordan, and by constructing a canal from the Mediterranean to the Dead Sea to take advantage of an even greater difference of level.

Lord Milk's recommendation appeared in his book *Palestine, the Promised Land*, first published in 1944, and caused great excitement in Zionist circles. A special society was formed to organise an advertising campaign to promote the project and to mobilise supporters in the United States. The report was taken as the starting-point for many other studies, some of which proposed considerable changes, but the general pattern of this project is still the dream of Zionists in their so-called promised land.

The Syrian-Jordanian Scheme for the Yarmuk River

A Syrian–Jordanian agreement concluded in June 1953 defined a plan for the exploitation of the waters of the Yarmuk river in which:

a. Syria was to use all the higher springs of the river in the area of Mezerili, the discharge of which in the summer is 135 million cubic metres. This project was carried out, and is known as the Mezerili Irrigation Project. The remainder of the Yarmuk waters were then to continue to flow on down to the Ghor lowlands in Jordan;

b. the two countries were to combine to build a dam on the Yarmuk river at Maqarin, for the storage of water and the generation of electric power;

c. four hydro-electric power stations were to be built between Maqarin and Adasiya, with Syria using 45 per cent of the power generated.

Estimated Average Yearly Flow

in millions of cubic metres (mcm)

Hasbani River at proposed site of Hasbani Dam	130
Jordan River below Lake Hula marshes (after correcting for marsh evaporation)	640
Flow into Lake Tiberias	838
Evaporation from Lake Tiberias	300
Net outflow from Lake Tiberias	538
Yarmuk River at junction with Jordan	475
Beisan Springs	67
Wadi Zarqa	45
Other inflow below Lake Tiberias	194
Flood flows developable	193

Source: Charles T. Main Inc., *The Unified Development of the Water Resources of the Jordan Valley Region*, Boston 1953.

For clarification, the first two figures give their portion of the supply which could be developed upstream of Lake Tiberias. The figures beyond that point from the Main Report estimated a developable water supply of 1,482 mcm. As it has turned out over the past years since the United Nations commissioned this report, rainfall in the Jordan Valley has been less than the averages given by Main, though in terms of long-term prediction it is thought that they represent measurable estimates of the amount of water available.

The Johnston Plan for the Jordan Valley

Following the announcement of the Syrian–Jordanian project for the Yarmuk river, an American scheme called 'The Unified Project for the Exploitation of the Water Resources of the Jordan Valley' suddenly came into existence. This project became known as the Johnston Plan, after Mr Alec Johnston, who presented it to the Arab countries on behalf of President Eisenhower. It provided for:

a. a storage dam to be built across the Hasbani river in Lebanon, linked by a canal connecting the dam with a hydro-electric power station at Tel Hej in occupied territory; the canal water would then be used for the irrigation of the Galilee area;

b. dams also to be built on the Baniyas and Dan rivers to provide water via a canal 120 kilometres long to irrigate the Galilee and Merj Ibn-Amer area;

c. the Hula swamp to be drained, accompanied by

d. the raising of the dam at the southern outlet of Lake Tiberias so as to increase the water level by about 2 metres and to bring its capacity up to about 300 million cubic metres;

e. a storage dam with a capacity of 175 million cubic metres to be constructed on the Yarmuk river at Maqarin;

f. a dam to be built on the Yarmuk river near Adasiya to divert its water into Lake Tiberias and to the East Ghor Canal;

g. two main canals to be dug on the east and west banks of the River Jordan, between its outlet from Lake Tiberias and its inlet into the Dead Sea, to irrigate the Ghor lowlands;

h. various other uses would be made of the water for such purposes as generating further electric power, controlling the flow of water in some of the valleys, and distributing the water fairly among users of irrigation.

The Johnston project allotted 45 million cubic metres to irrigate 3,000 hectares in Syria; 744 million cubic metres to irrigate 50,000 hectares in Jordan; and 394 million cubic metres to irrigate 42,000 hectares in Israel. It became clear that this project did not give the Arab countries their rightful share in the development of the waters of the River Jordan and its tributaries, so the Arab League appointed a Technical Committee of Arab engineers to study the Johnston plan. This committee reached the following conclusions:

a. The project disregarded the political frontiers of countries whose territory lay in part in the basin of the River Jordan. This point is illustrated by the facts that though a dam was to be built on the Hasbani river in the Lebanese part of the basin, Lebanon would derive no benefit from it because all its water would go to Israel; nor would Lebanon derive any benefit from the hydro-electric power station which was to be set up in Tel Hej in the occupied territory but which would be driven by water brought by a canal originating in Lebanon; the plan to divert the Baniyas river into occupied territory would bring no benefit to Syria, and the plan proposed to divert large amounts of water from the Yarmuk river to be stored in Lake Tiberias, thus placing the Jordan lowlands at the mercy of Israel, albeit under the supervision of an international committee.

b. The Johnston plan would increase the salinity of the water from the Yarmuk river when it mingled with that of Lake Tiberias, the salinity of which is higher than that of the river.

c. The proposed increase in the surface area of Lake Tiberias would result in a greater loss by evaporation.

d. The proposed distribution of water among the Arab countries concerned would be unfair. For example, an area of 3,000 hectares in Lebanon would not be irrigated by water from the Hasbani river, and only 45 million cubic metres were allotted to an area of 3,000 hectares in Syria, where there would still remain an area of some 13,000 hectares in need of irrigation and requiring 132 million cubic metres from the Yarmuk river.

e. The amount of water allotted to Jordan under this project would not suffice for its present needs and for the future requirements of the agricultural schemes which form so essential a part of her economic development programme.

f. The project as a whole allotted only 819 million cubic metres of water to the Arab countries, whereas the total amount flowing from these countries into the River Jordan was estimated at about 1,283 million cubic metres. The Arab economies were thus to be the losers by some 464 million cubic metres.

In the light of these observations, the Arab Technical Committee met in March 1954 and prepared an Arab counter-project, the main points of which can be summed up as follows:

a. The building of a storage dam with a capacity of 60 million cubic metres on the Hasbani river at a point inside Lebanon about 20 kilometres from its confluence with the River Jordan, thus making possible the irrigation of 3,000 hectares of Lebanese land within the river basin. This area would require 30 million cubic metres of water, leaving the remainder to run into the River Jordan. A hydro-electric power station would also be constructed on the lower stretch of the Hasbani near the Lebanese–Syrian border.

b. The construction of a diversion dam on the Baniyas river in Syria, channelling the water by two canals to the east and west to irrigate an area of nearly 2,200 hectares in Baniyas, the remainder again being left to flow into the River Jordan.

c. As with the other projects put forward, the building of a storage dam on the Yarmuk river, on a site at Maqarin, as well as a diversion dam at the Adasiya site. A hydro-electric power station would also be built here, the water from which, amounting to 60 million cubic metres, would be stored in Lake Tiberias for use in the irrigation of the lowlands of the Jordan Valley.

d. The construction of a canal on the West Bank of the River Jordan to irrigate the western lowlands.

e. The formation of an international organisation to supervise the various elements of the project and to ensure a fair distribution of the water.

The Arab League discussed their proposals with Mr Johnston, who agreed with some of the points but objected to others. His most important objection concerned the amount of water allotted to each area of land, the amounts defined in the Arab project being 15 per cent more than those provided for in his project. Johnston insisted on allowing 394 million cubic metres of water to Israel, as outlined in his project. He was determined that Israel should not be restricted to using only its contributed share in the river basin but should be allowed to take any amount it needed, while the representatives of the Arab countries were equally insistent that the case they had put forward was a just one, bearing in mind the likelihood that Israel would transfer vast amounts of water to the Negev Desert.

The Israeli Project

The Israeli authorities themselves prepared several schemes to seize the waters of the River Jordan, disregarding the rights of the Arab countries to use this vital source of natural wealth. In 1954 they started work on such aspects as:

a. The draining of Lake Hula, begun in 1951 and completed in 1956. This helped to restore a certain amount of water previously lost by evaporation.

b. The construction of a dam on the River Jordan at Kasr Attra so as to divert its waters by means of a canal which flows into Lake Tiberias at Tayffa. This canal had been commenced in 1951 but was stopped as a result of the Syrian Government's protest, supported in 1953 by the UN Security Council's Resolution.

c. The construction of channels, tunnels and dams, and the installation of pipes and pumping stations to draw water from Lake Tiberias, and from the Boutoun basin in particular, for use in the Negev Desert, have gone ahead and the Tiberias–Negev conduit is already completed. All indications point to the technical success of this experiment.

The Israeli project was based on the idea of making Lake Tiberias the main storage centre of the waters of the Jordan, after raising the level of the lake by $1\frac{1}{2}$ metres, and diverting some 320 million cubic metres annually to the Negev Desert, which lies wholly outside the Jordan basin.

At present, Israel benefits from more than 100 million cubic metres of Jordan water, so that now all the potentially arable land is being cultivated. In 1967, 162,000 hectares were under irrigation out of a total area of two million hectares.

The disadvantages of the Israeli project from the point of view of the Arab countries were that:

a. Israel would benefit from the whole of the Jordan waters before the river entered Jordanian territory. Moreover, by the diversion of the saline springs to the river, the waters in that stretch would become so saline as to be useless for irrigation or other purposes, thus causing great damage to Jordan, especially to the 7,000 hectares in its lowland area which depend entirely on irrigation water pumped from the river.

b. The Israeli project would constitute a flagrant violation of the rights of the Arab countries, parts of whose territory lie in the Jordan basin and who have a right to benefit from its waters.

c. Israel was endeavouring to open major new agricultural horizons in the occupied territories so that she could consolidate her position and continue her defiance of the Arab countries.

d. Finally, the effects will show in a gradual lowering of the level of the Dead Sea, increasing the problems of salinity.

The New Arab Project

Immediately following the publication of the Israelis' plans for the diversion of the waters of the River Jordan and its tributaries into the occupied territories and the initial steps taken to put it into effect, the Arab countries protested against this project and threatened to use force if Israel did not stop carrying it out. Meanwhile, the Israeli authorities were preparing themselves to defend and protect it by all possible means, even by resorting to war, claiming that it was a matter of life and death to Israel. Since Jordan was the country most harmed by the implementation of the proposals, she herself presented a counter-project in November 1960, which aimed at diverting the waters of the Hasbani and Baniyas rivers into Jordanian territory and thus preventing it from reaching the occupied areas. This came to be known as the New Arab Project. Its main objectives can be summarised as follows:

a. The building of a storage dam on the Hasbani river and the diversion of the water surplus to Lebanon's needs to the River Baniyas by means of an open canal; and in addition, the diversion of the waters of the Bareighit river in Lebanon into the Hasbani Canal near Baniyas.

b. The construction of a diversion dam on the River Baniyas at an altitude of 380 metres above sea-level, and the provision of an open canal to pass across the south-western region of Syria. This canal would receive the surplus waters of the Hasbani, Bareighit and Baniyas rivers and lead them into the Rakad Valley.

c. The building of a storage dam at Nahila to hold the accumulated water diverted from Lebanon and Syria.

d. Finally, the implementation of the joint Syrian–Jordanian project for the exploitation of the Yarmuk waters, as provided for in the agreement concluded between the two countries in 1953.

This was adopted by the Arab League. Foremost among its advantages was that it would ensure Jordan's requirements of water to irrigate the whole of the Ghor lowlands and also meet the needs of Syria and Lebanon.

The Joint Arab Project

Following the implementation of the Israeli project and the realisation of its serious consequences, the Arab countries rose in unison to confront this grave situation. An Arab Summit Conference was convened in Cairo in January 1964, which authorised the Secretary-General of the Arab League to appoint an Arab Technical Committee to study the subject and to arrive at a joint Arab project to counter that of the Israelis. The Technical Committee, consisting of representatives of the Arab countries, considered the following points:

a. a survey of the latest stage reached by the Israeli project in diverting the Jordan waters;

b. the material damage inflicted on the Arab countries by its implementation;

c. what positive steps could be taken to utilise the waters of the tributaries of the Jordan that lay within the Arab countries.

With regard to this last point, the Technical Committee drew up plans which the Arab Summit Conference approved. This provided for the exploitation of the Hasbani river, including the Baniyas and Yarmuk springs. In spite of the amount of preparation and detailed research which such a project necessitated, the Committee considered it most important that it should be carried out as quickly as possible, to enable the Arab countries to counter the danger threatening their interests as a whole. Consequently it divided the project into two phases, the immediate and the longer-term.

The immediate measures, which were given first priority, involved works to be undertaken in the Lebanon and Syria, already described.

In addition to the measures undertaken in Lebanon and Syria, the Jordanian Government financed the work immediately necessary to raise the level of the eastern Ghor Canal so that the surplus water from the Wazani spring and the Baniyas river could reach the upper area of the Jordanian Ghor, which was to suffer most from the Israeli project.

The immediate measures recommended by the Technical Committee and outlined above will guarantee the exploitation by the Arab countries lying in the Jordan basin of about 230 million cubic metres of water annually; that is, about 72 per cent of the amount Israel draws from Lake Tiberias and pumps into the

Negev Desert. After Syria and Lebanon have secured their needs from the Wazani–Baniyas–Yarmuk Canal, Jordan will in turn benefit from a part of the waters of the River Jordan and its tributaries by the diversion of its waters from Israel—waters of which Israel intended to deprive the Arab countries.

As for the long-term plans, the Arab Technical Committee found it necessary to defer this part of the work until after the immediate measures were completed. It then proceeded to draw up plans for storage schemes in different areas, particularly in the Hasbani region. These schemes have also been embodied in the Joint Arab Project for the exploitation of the Jordan waters. It was agreed that a special body to be called the Organisation for the Exploitation of the Waters of the River Jordan and its Tributaries should be set up, with its headquarters in Cairo. The Organisation is affiliated to the Arab League and assumes direct supervision over the planning and implementation of its works. It is administered by a Board of Directors presided over by the Secretary-General of the Arab League. Its members include representatives from Lebanon, Syria, Jordan and Egypt, and it is supported by administrative and technical services under a Director-General.

Responsibility for the implementation of the project lies with three executive organs, namely the General Projects Organisation in Syria, the Litani Administration in Lebanon, and the Jordanian Rehabilitation Council in Jordan. The Organisation's Board of Directors meets monthly in each of the four Arab countries to follow up the work carried out by the three executive organs. Work is proceeding smoothly, rapidly and fully according to plan. The Board of Directors have introduced certain technical modifications to the immediate measures, without changing any of its main objectives or priorities, which still are: first, to construct the canal to divert water from the Serlid spring in Lebanon into the Wazani–Baniyas Canal; second, to increase the capacity of the Baniyas–Yarmuk Canal from 7 to 12 cubic metres per second; and third, to construct a storage dam at Mokhiba to hold the water coming from the Baniyas–Yarmuk Canal in winter.

In conclusion, it is perhaps worth observing that the reasons why so many projects for the exploitation of the waters of the River Jordan and its tributaries have made their appearance are, first, the very complex and highly interlinked physical and political nature of the river basin, and second, the necessity for the Arab countries to take measures to repel the aggressive intentions of the Israelis.

REFERENCES

Alice Taylor (Ed.), *Middle East*, David and Charles, 1972

The Middle East and North Africa 1973-74, 20th edition, Europa Publications

Michael Adams (Ed.), *The Middle East : A Handbook*, Anthony Blond, 1971

J. I. Clarke and W. B. Fisher, *Populations of the Middle East and North Africa : A Geographical Approach*, University of London Press, 1972

Peter Mansfield, *The Middle East : A Political and Economic Survey*, 4th edition, Oxford University Press, 1972

The Republic of Iraq

Like Egypt, Iraq is one of the first countries in history where the people settled, cultivated the land, and established a great and lasting civilisation based on agriculture. It became an independent state in 1932, and is luckier than many new nations in that it has petroleum, a large source of capital for not only improving technological development but for long-range projects such as flood control and water storage, administered by the Ministry of Planning.

The country occupies an area of some 436,000 square kilometres, and its ancient name of Mesopotamia—the land between the rivers—indicates that it is a country largely dominated by two of the world's greatest rivers: the Tigris (1,850 kilometres long), and the Euphrates (2,650 kilometres), both of which rise in eastern Turkey. They form a pair of arteries, running through Iraq from north-west to south-east, merging to form a single stream, the Shatt al-Arab, at Qurna, 185 kilometres from the sea. To the west and south of this vast river-valley, the desert plateau rises gradually to merge with that of Syria, Jordan and Saudi Arabia. The frontiers with these countries are artificially-drawn lines. In the east the sharp wall of the Zagros Mountains of Iran rises to 3,000 metres, while in the north, Iraq is bounded by Turkey, and in the south-east by a few kilometres of coastline at the head of the Arabian Gulf and by the State of Kuwait.

The population of about 9·5 million is spread over the total area of the country but nearly 50 per cent are now concentrated in urban areas. Fewer people are engaged in agriculture than was the case ten years ago, but in spite of this movement from rural to urban areas (a general trend) over some years, agriculture still occupies more people than any other sector. Agricultural exports (oil excepted) comprise about three-quarters of the total. Natural resources and possibilities for development differ widely throughout the country because of variations in climate, soil, and transport conditions.

Agriculture in Iraq depends on both rainfall and irrigation. The irrigated lands lie in the alluvial plain in the central and southern parts of the country, while the areas depending on rainfall are in the mountainous and semi-mountainous regions to the north, though in these regions too there are some areas which depend on irrigation from river and spring water.

AREAS OF AGRICULTURAL PRODUCTION

These areas fall naturally into two regions: a rainfall area and an arid but irrigated area. Both factors play an important part in relation to the crops grown, and most significantly form the boundary of the northern limit of the date palm. The farmers cultivate those pieces of land which are the most fertile and which enjoy abundant water, leaving extensive areas uncultivated and neglected because of poor soil or inadequate supplies of water. This means that in the south barley, rice and dates are grown as staples under irrigation; in the north, barley, wheat and fruit are the major crops.

In addition to the difficulties caused by the climate, rivers with continually shifting banks, marshland, and rapid floods, all manner of insect pests have in the past made high agricultural yields difficult.

Generally speaking, the agriculturally productive areas of Iraq are sparse, and are separated from each other by extensive uncultivated areas. Some of these are cultivable but are left fallow to be exploited during the next season, while some are cultivable but are left unexploited.

The most important areas of agricultural production are:

The Shatt al-Arab

This area extends on both sides of the Shatt al-Arab River between the confluence of the Tigris and Euphrates at Qurna and the mouth of the conjoint river at Faw. The area possesses an abundance of water, irrigated as it is by the tidal ebb and flow of the Shatt al-Arab and by the streams flowing into it on both sides. This tidal irrigation of fresh water is very easy and does not need much manpower. The area is considered to be one of the best in the world for date palm orchards (Iraq's major export crop) in spite of the high salinity of its soil.

The name *Iraq*, which in Arabic means cliff, is descriptive of much of the topography of the land.

An established irrigation system on the River Tigris, as seen from the air.

The Amara

This includes the land situated on both banks of the Tigris between Qurna and Amara. There are many irrigation schemes here and it is famous for rice and millet, which are grown in the marshes, and for wheat and barley, which are cultivated on the higher land near the river.

The Gharraf

This area, extending on both banks of the Shatt al-Gharraf, an old drainage channel between Kut on the Tigris and Nasiriya on the Euphrates, was once one of the most fertile agricultural areas in Iraq and in ancient times was famous for its wheat and barley. Today, because of the Kut barrage on the Tigris and the introduction of better strains of barley, mostly from Morocco and California, which are more tolerant of salinity, cultivation of the area by irrigation has been made possible, and in this way the region has regained much of its former importance. Iraq now exports barley.

The Ash-Shatra drainage project associated with the Gharraf scheme should increase cultivation by some 70,000 hectares.

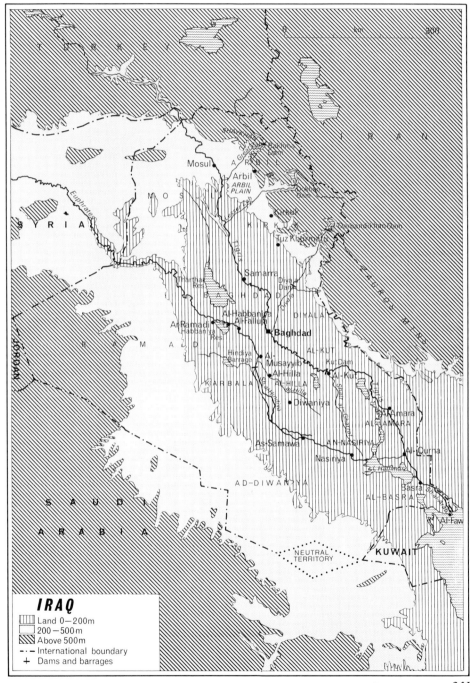

IRAQ

⊞	Land 0—200m
☐	200—500m
▨	Above 500m
–·–·–	International boundary
+	Dams and barrages

The Middle Euphrates

The area situated between the Shatt al-Hindiya and the Shatt al-Halma and lying between Musayyib and Samarra, where there are many channels and streams along both banks of the Euphrates, is one of the most productive in Iraq. It is the most densely populated agricultural region in the country. Rice is grown in the southern parts, and wheat and barley in northern area between Diwaniya and Hilla, while stock-raising also takes place in all these regions affected by the irrigation projects (see map).

The Middle Tigris

The land along both banks of the Tigris between Kut and Samarra was long neglected until barrages were constructed, because it rises well above sea-level. Wheat and barley are now grown in winter, and cotton, sesame and maize in summer.

The Baghdad Area

The area between the Tigris and the Euphrates near Baghdad, where the two rivers come close to each other, is an alluvial plain in which wheat, barley and cotton are the principal crops.

The Ramadi Area

Here the construction of the barrage has increased the area of cultivated land on both banks of the Tigris. Canals, aided by pumps and water-wheels, bring water to larger areas for greater crop cultivation.

The Mountainous Belt

This crescent of mountain countryside lies in the north-eastern part of Iraq, where the rainfall is adequate to support cultivation without irrigation, although there are places in the foothills where springs and artesian wells are used for irrigating small areas of land for cultivating fruit, rice, vegetables and tobacco. There are also patches of terraced cultivation (vineyards), with farmland scattered among the mountain valleys where rice and other grain crops are grown. Scrub oak covers much of the mountainsides.

This area includes:

The Diyala

Situated on the banks of the Diyala River, a tributary on the east bank of the Tigris, this area has increased its importance since the building of the Darbandi-khan Dam in 1940. More than 400,000 hectares are watered by flash irrigation from the many streams which receive water from the main river channels in this area. The region is the most important fruit-growing area in Iraq; citrus fruits, vines and pomegranates are plentiful, while wheat and barley are grown in winter, and cotton, rice and tobacco in summer.

The Zab Area

This large tract of uncultivable desert includes much of the Mosul plain between the Tigris and the mountainous areas of Shaykhan and Sengar; coupled with the Eski-Kalek project on the left bank of the Greater Zab River, it also takes in the Arbil plain, which lies between the Greater and Lesser Zab rivers. If the project for the Dibbis Dam on the Lesser Zab is completed, this part of the government's scheme should irrigate the Kirkuk plain between the river and the town of Tuz Khurmatu. Then the picture would be changed: some 12,000 hectares would be irrigated, whereas at present the whole region depends on rainfall (approximately 330 mm) for the cultivation of various kinds of winter grain.

Salt crusts formed along irrigation ditches between wheat fields.

Water melons

Afforestation on the mountainous hillsides at Ain Zala shows one method of halting soil erosion, and is part of the Government's long-range programme.

Quinces (right).

On an experimental farm near Baghdad the Friesian has been introduced into a scheme for rearing various breeds of cattle.

The problem of soil salinity is always present where irrigation is practised. Sugar cane appears to be one of the crops least affected by a high salt content.

AGRICULTURAL DEVELOPMENT

During the past ten years there has been considerable progress in the agricultural sector. Providing the dams, reservoirs, and artesian wells are completed, and not held up or left at the planning stage, the area of cultivated land in Iraq could be twice what it is at present; in addition, such constructions would help to protect the Iraqis from floods, and could provide sufficient electric power for the whole country.

As it is, with extensive pastoral areas at present beyond the possibility of irrigation, the farmer clings to his ancient methods: tools and implements he inherited from his forefathers, the wooden plough drawn by draught animals for tilling the soil, the hand-scythe for harvesting the crops, the wooden thresher for husking his grain, and those most primitive aids of all, his own muscles and those of his animals, in raising irrigation water to his land. Nor can he afford to buy chemical fertilisers except on a very limited scale. However, the co-operative societies have begun distributing fertilisers to the farmers, who pay for them after harvest time. Happily, Iraq can herself produce artificial fertilisers and other chemical products, because the necessary raw materials are available within the country and can be economically extracted and processed.

LAND-OWNERSHIP

For a long period after the Mongols brought about the downfall of the Abbasid State in the thirteenth century, disorder and confusion prevailed in Iraq. Bloody wars were waged between the settled tribes and the bands of nomads coming from the desert in search of fertile pastures. Gradually, these various tribes began to join forces and combined together in order to settle in certain areas, where they eventually established a claim to ownership of the land. In this way, large tribal groups in the form of independent emirates came into being and often fought each other in dispute over the land.

Outside the towns, the central authority had little power, and a primitive tribal system maintained itself for a long time, developing towards feudalism as the power of the tribal chiefs increased. A tribe's area of land, usually called the *dira*, was nominally owned collectively, not by any individual, and all had rights over the pasture and arable land; but in practice it was divided among the families of the tribe, and was further divided by the head of the family into small individual plots or strips. In return for his leadership and organisation, the head received a fixed share of the produce.

In some tribes the allocation of the land was permanent, in that each family continued to exploit its own plot or plots by succession from father to son. In other tribes the occupancy of the different plots was changed by rotation, the land being re-divided among the families after periods ranging from one to four years.

The members of a family were free to farm their land or to rent it out to others in the same tribe in return for a certain percentage of the produce, usually between a third and a half of the yield. The net revenue of the farmer, after paying out the shares of the owner and supervisor—the head of the family or the tribe's sheikh—ranged between a fifth and a third, but was sometimes as much as half, of the produce.

This was the system generally practised in Iraq until the beginning of this century. But, even by the middle of the last century, fundamental changes were beginning to take place, producing a trend which resulted in concentrating land-ownerships in a manner that had far-reaching effects on the economic, social and political life of the country.

The Ottoman government had applied to land-ownership the rules of Islamic jurisprudence, without keeping any register or records. In an attempt to remedy the long-standing confusion which had resulted, the government enacted a land law in 1858, defining, fixing, and registering the ownership of land. Besides its overt purpose, however, this law was also intended to weaken the power of the tribal sheikhs by permitting individual ownership of small areas, and to facilitate the collection of taxes which, under the tribal system, had been withheld with impunity. Consequently, the new law produced results contrary to its nominal intention, and the ownership of land became concentrated rather than distributed.

The British occupation, which succeeded the Ottoman rule, preferred to sustain the tribal system as a basis for support. Large holdings of land were given to tribal sheikhs and to townsmen, and a feudal class developed which expropriated tribal lands and, often by supporting the British authorities, enormously increased its own power and possessions.

At the beginning of independence in Iraq, most of the land records were lost, and the government re-introduced the old Ottoman Land Law. This had no greater success than before, and the services of a British expert, Sir Ernest Dawson, were enlisted. In 1931 he published his report, on which was based a new Settlement Law. Whatever may have been the purposes of this legislation, the result was that by the nineteen-fifties a small number of tribal sheikhs owned vast areas of land, while the majority of farmers eked out a scanty livelihood. The statistics for 1952 show that 125,000 individuals owned 6 million hectares, 27 per cent of the whole country. The average area of individual ownership varied from one province to another, from some 6 hectares in al-Basra to 1,700 hectares in al-Amara province. A very small number of landowners owned estates of 120,000 hectares.

These average figures alone cannot give a true picture of the maldistribution of land in Iraq at that time. A detailed analysis shows that 24,270 persons owned less than a hectare each; 25,849 owned between 10 and 20 hectares; 33,021 owned more than 100 hectares. About 85,000 owners, 68 per cent of the total, held only 8 per cent of the privately-owned land, while 39 per cent owned 85 per cent. There were, of course, very many people working on the land who owned none of it.

Apart from land owned privately or under entailed rights, 66 per cent of the area of Iraq was owned by the state; the exploiters of this land acquired the full right of disposing of it and enjoyed all the benefits from it. The rights varied according to the type of exploitation. *Tabu* land was for all practical purposes owned outright, although the government had in theory the right to repossess it if the exploiter neglected it for three successive years. Of all agricultural land in Iraq, 40 per cent was *tabu* land.

In some parts of the country, a system of land-concessions obtained, whereby the sheikhs were given a concession for an area in return for a sum payable by instalments. This system was harmful to the interests of small farmers. A further form of land-concession involved the leasing of state-owned land to individual farmers.

The whole problem of land-ownership was greatly complicated by the fact that land was unsurveyed and title-deeds were inaccurate. The government enacted a number of laws to remedy this, between 1932 and 1940, but there were still vast areas unsettled eighteen years later. In 1945 the position was improved by the development of state-owned land and its distribution to cultivators; but this had no application to private estates, so that only a few farmers received any benefit, and vast areas were left in the hands of the big landowners.

Such were the conditions prevailing in Iraq at the time of the Revolution of July 1958. In September 1958 the new revolutionary government promulgated the Land Reform Law to Eliminate Feudalism, which determined land-ownership and distributed land among the farmers, to raise their standard of living and to increase the national income. The law fixed a maximum holding of 370 hectares in irrigated areas and 500 hectares in unwatered lands. The government was to requisition the surplus areas, with adequate compensation, and redistribute it, together with state-owned lands, among small farmers, in plots of from 7·5 to 15 hectares in irrigated lands and from 12 to 30 hectares in unwatered lands, at a purchase price payable by instalments over twenty years.

The law organised farmers into co-operative societies, with the function of supplying farmers with seeds, fertilizers, and agricultural implements, and helping them in financing crops and marketing produce. The societies also removed the old feudal ties, ensured justice between landlord and tenant, and guaranteed the rights of agricultural workers. These model schemes are run rather like the co-operative scheme in the Gezira of the Sudan: overall assistance is given, not only by means of credit, but including marketing and other technical help. The unit system of granting land was based on the environmental resources such as irrigated land, land watered naturally by free-flow, land with adequate rainfall, mountainous areas, and swamps. Maximum holdings were 750 hectares of rain-fed land, 500 hectares of irrigated land.

The first co-operative scheme was at Dujaila, near Kut. Here over 1,000 plots each of 24 hectares were allotted to farmers for farming under the supervision of the government.

Intensive farming methods were used without the usual practice of allowing the land to lie fallow, and this involved the planting of three crops rather than the more general two. In this way, production is increased and the income of the farmer raised.

A similar project under way in the Musayyib, when completed, should form sixty villages covering about 50,000 hectares and giving agrarian work to 35,000 people. Some encouragement is thus given for the rural migrants to return from the cities to the land again.

By improving strains of livestock, varying the types of crops, reafforestation of the hillsides in the north, and continuing to improve drainage and fallow crops in order to reduce the problem of salinisation, Iraq is establishing a sound base for future prosperity.

WATER

The average quantity of sediment in the water flowing down the Tigris River in flood seasons is as much as 2,300 grammes per cubic metre while that of the Nile in flood is 1,700 grammes. The graph shows the seasonal rate of flow of both the Tigris and the Euphrates, indicating that the flow fluctuates considerably from one season to another, reaching its lowest during late summer. The two rivers are usually in flood by the end of the winter and in early spring after the winter rain and the melting of the snows on the mountain heights of Turkey, Iran and northern Iraq. Because there is a forceful current in both rivers, erosive action is considerable, and a great deal of sediment is carried at all seasons. This contrasts greatly with the Nile, which for seven or eight months of the year is clear and quiet, while the rivers Tigris and Euphrates are turbid and rapid at all times. The water-level in the two rivers falls so low in summer that crops are often destroyed, yet when they are in flood an enormous volume of water spills out into the southern marshes or flows uselessly into the Arabian Gulf. The Euphrates has two notable left-bank tributaries, the Balikh and the Khabur; the latter, flowing across the hot plains of Syria,

Irrigation projects have greatly increased the potential agricultural yield in Iraq.

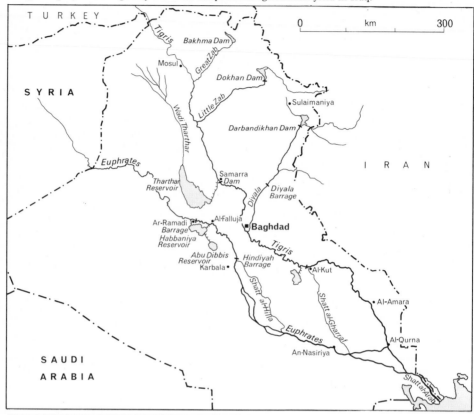

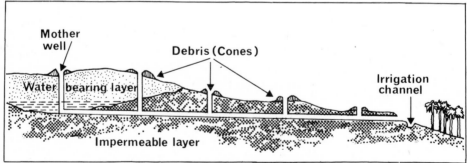

A *falaj* (plural *aflaj*) known in Iraq as *karez* or *foggara*, is an ingenious method of irrigation which has survived for centuries. Water is trapped between two layers of rock. A sloping tunnel is dug, lined with brick, stone or tiles, which drains off the trapped water at a lower level. As in a mine, vertical shafts are dug so that the debris from the tunnel can be brought to the surface. The site of a *falaj* is littered at intervals with cones made from the debris, which resemble giant mole-hills. The water channel can often be extensive, some 50 kilometres or longer. Maintenance is of prime importance to avoid the *falaj* falling into decay.

is subject to considerable evaporation, which reduces the volume of water—a problem similar to that of the Nile.

In the northern areas which depend primarily on rainfall, summer crops are not extensively cultivated because of the absence of rain in summer. The area put under summer crops is less than 20 per cent of that under winter crops. This explains why Iraq devotes only limited areas to growing cotton and rice. Cotton is

A typical monthly flow pattern of the Tigris and the Euphrates Rivers.

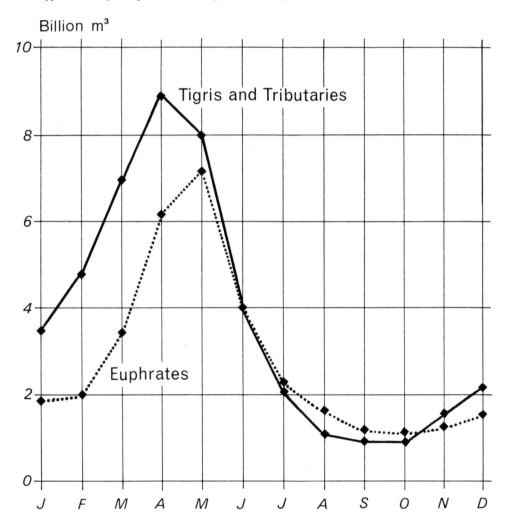

particularly susceptible to insect pests and soil salinity, which is prevalent where irrigation is practised. Also, when cotton most needs water, in July and August, water is scarce and the climate extremely hot and dry; and where rice, which is a main summer crop in Iraq, should be widely grown its cultivation is limited by scarcity of water and the need to save what there is to irrigate orchards.

Downstream of a head regulator leading to an earthen canal. Water is conveyed to the land, in this instance for an afforestation scheme, by gravity flow. Some of the problems such as weed growth and silting illustrated here show the need for continuous maintenance. This method of irrigation is very common in central and southern Iraq.

IRRIGATION

Just as Egypt is the gift of the Nile, so is Iraq the gift of the Tigris and the Euphrates. Without them, the fertile plain between would have been a barren desert, for although about half the agricultural land receives adequate rainfall, in the northern areas the remainder relies on seasonal irrigation from these two rivers, which are the mainstay of free-flow irrigation.

The control of irrigation by river waters is distinctly different, and is a problem that might be described as unique to Iraq. Simply because of the timing of the floods, which occur in spring when crops are only partly grown (rice being the exception), the inundation of cultivated areas could be disastrous. Thus the water of the rivers has to be kept within the banks by the construction of artificial embankments, which can be breached if necessary at selected places should flooding occur. In addition, the drainage of stagnant flood water is paramount, because if retained it tends to induce soil salinity. This badly affects such crops as wheat and cotton, although the date palm can stand up to it better.

The rivers of Iraq, like the River Nile, have flood seasons and dry seasons. During flood seasons, as we have seen, water flows in great quantities to the sea and the lakes without being put to much use. The floods often cause great damage when the flow of water suddenly becomes abnormal and inundates extensive areas of land. (Baghdad was badly flooded in 1954 and again in 1958.) The control of floods and the utilisation of river water for irrigation are therefore very important to Iraq's agriculture, as is also the containment of surplus water in the flood season for use in the dry season.

Egypt makes use of the flood waters of the Nile—and did so long before embarking upon modern irrigation projects. The rivers of Iraq are more complex: the river system has extreme annual variation in flow, sometimes causing years of drought and years of flood. While the Tigris (the larger of the two) reaches its maximum level in April, the Euphrates does so later, which means that the flood season comes too early for maturing summer crops and too late to help winter cultivation.

For these reasons, successive governments that have ruled Iraq have given considerable attention to irrigation and agricultural methods. History records that the Sumerians, Akkadians and Amorites showed great skill in digging canals for irrigation purposes. The Persians and the Arabs were also skilful in this field. There were famous canals in the Persian and early Arab eras, such as the Nahrwan Canal which irrigated the land lying on the east bank of the Tigris between Samarra and Kut and which continued to flow until the twelfth century. The ancient course of that canal is still visible, dug deep into the earth from north to south along the Tigris River. A succession of conquerors were content to leave these canals alone, all wisely realising that once the waters were out of control they could well become their master. The Arab State in Iraq continued to take good care of irrigation systems until it was invaded by the Tartars in the thirteenth century; these were nomadic tribes who neglected irrigation and agriculture with the result that the canals silted up and the earthworks on the banks of the rivers were destroyed, reducing the land to a wilderness of scrub and marsh. With the advent of Ottoman rule in Iraq, thought was given to controlling and benefiting from the river waters. The Turkish government invited an Egyptian engineering mission headed by Sir William Willcocks to report on flood control and irrigation. In 1911 this mission presented its report, the principles of which were put into effect half a century later. The Turkish government carried out some of the projects proposed, the most important of which was the Hindiya Barrage on the Euphrates, and the control of flooding of the Tigris.

The broad planning of the irrigation system evolved round the variation in level between the Tigris and the Euphrates. The bed of the Tigris lies slightly higher than that of the Euphrates from Baghdad to Qurna, which makes the construction of free-flow irrigation channels an easy task. In the Baghdad region, however, the reverse is true, with the Euphrates lying higher than the Tigris. These two facts have been exploited by irrigators for centuries.

Above Ramadi both rivers flow in well-defined valleys comparable with the Nile. The Tigris meanders between natural banks making the entry of tributaries difficult—a fact peculiar to this river.

Therefore the irrigation projects in Iraq have been planned to include three different types of control: namely, reservoirs, barrages, and canals.

The Reservoirs

There are two kinds of reservoir in Iraq. The first consists of those established in the neighbouring desert depressions in which flood waters are stored for use in times of drought. Among these are the Habbaniya and Abu Dibbis basins and the Tharthar Reservoir. The Abu Dibbis Depression is at present used for emergency measures to take excessive flood water from Lake Habbaniya, and is not used for irrigation, as Abu Dibbis water has extensive deposits of salt. The second kind comprises those established on the course of the river itself to store the river-waters. Among these there are the Dokhan and Darbandikhan on tributaries of the Tigris.

The Desert Depressions

The Habbaniya Depression is situated south of Ramadi 5 kilometres from the west bank of the Euphrates and is about 24 kilometres long and 12 kilometres wide. To the south there is Abu Dibbis which is about 30 metres below the flood-water level at Ramadi. In 1911 the Willcocks report recommended using the Habbaniya Depression and the Abu Dibbis for storing the waters of the Euphrates. This project was carried out in 1951 and included the following:

a) A barrage across the Euphrates at Ramadi; this has twenty-four openings, with a navigation lock for river craft and a fish ladder.

b) An inlet channel called the Warrar, which raises the river level to carry the waters to Lake Habbaniya. The lake has been banked to increase its storage capacity.

c) An outlet canal at Dhibban to carry the water from Habbaniya back into the Euphrates during the dry season. This canal flows into the Euphrates at Fallujah and there is a barrage at its outlet to control the water discharge.

d) An escape regulator at Mujara on the south bank of the lake, to cope with excess floods. This discharges water into the Abu Dibbis Depression in the desert, from which it is then lost by evaporation.

The capacity of the Habbaniya Reservoir is 2·3 milliard cubic metres of water which has been increased to 3·2 milliard by heightening the Ramadi Barrage. The capacity of the Abu Dibbis is about 11 milliard cubic metres of water. Thus the Habbaniya Reservoir serves a double purpose: it lessens the danger from severe floods on the Euphrates and stores considerable quantities of water for use in agricultural expansion projects. Smaller schemes on the Euphrates such as the Greater Musayyib project have irrigated some 100,000 hectares. In addition, regulators in the lower part of the river before Lake Hammar and at Shamiya have done much to maintain river level during the period of the dry season.

Iraq's largest flood-control scheme is situated in the Wadi Tharthar on the Tigris. Because of the greater volume of water, the basin at its lowest point, three metres below sea-level, is similar in size to the Dead Sea. As is common in many parts of the Middle East where rainfall is scanty and evaporation rapid, drainage takes place within the enclosed area of the basin, this type of inland drainage being known as aretic.

A barrage has also been built at Samarra to raise the level of the water above it, and to deflect the waters through a canal some 65 kilometres long to the Wadi Tharthar storage dam, which has a capacity of 8·3 milliard cubic metres of water. The fall in the level is ideal for the generation of hydro-electric power. This scheme not only helps to protect the area from the dangers of the Tigris floods, but approximately 190,000 hectares of agricultural land are now irrigated by it. As flood prevention it has proved most successful.

The Dokhan Dam

This is another regulator dam, built on the Lesser Zab River at the Dokhan Gorge. It is a high dam (105 metres), and in normal conditions the capacity of the reservoir above it is 4·5 milliard cubic metres of water. The dam controls the flood waters of the Lesser Zab River and stores water for irrigation, distributing it to a wide area of land, estimated at over 300,000 hectares, in the plains lying between the Lesser Zab and the Greater Zab, both of which dry up in summer. It is also a source of hydro-electric power.

The Darbandikhan Dam

This is a multi-purpose dam built on the Diyala River, the most important tributary of the Tigris, at a point 60 kilometres south-west of the town of Sulaimaniya. The dam, which was built in 1960, helps to protect the area against the danger of severe floods, and irrigates an area of 125,000 hectares.

Other schemes such as the Bakhma Dam at the Bakhma Gorge on the Greater Zab River, and another on the Tigris north of Mosul, are fully operational.

265

The Control Barrages

Many older barrages and weirs are designed to raise the water-level and to control the distribution of water in the canals emanating from them.

The oldest of these is at Hindiya, and was constructed between 1912 and 1915 on the Euphrates above its confluence with the Hilla River. It was built in the Egyptian style and has thirty-six openings each nearly five metres wide. As there were no quarries near the site, it was made of bricks. Although the barrage was designed by Sir William Willcocks, one of the most famous irrigation engineers at the time, it has many defects. It diverts water from the Euphrates into a canal taking a similar course to the 'new' river through Hilla. By raising the level of the Euphrates it makes irrigation possible over a wide area in the regions of Karbala and Ramadi, as much as 150,000 hectares. If more canals can be provided, there is a potential irrigation area of some seven million hectares.

The Kut Barrage, more than three times the size of the Hindiya Dam, was constructed in 1943 on the Tigris at Kut, to make possible the cultivation of two large areas, namely the Shatt al-Gharraf on the west bank of the river, and the Shatt al-Dujaila Canal on the east bank. It is 492 metres long and has fifty-six openings. Nevertheless it is of seasonal value only, as the summer flow is required for rice cultivation at Amara, farther downstream, and only the winter supply can be used for these areas.

The Diyala Barrage feeds six canals for the irrigation of extensive areas both north and north-east of the city of Baghdad.

Inauguration of the Hilla Canal in 1913.

The Canals

The development of modern irrigation is associated with the digging of a wide network of canals throughout the country. A canal has already been made in the Kirkuk Plain which draws water from the Lesser Zab River to irrigate a further 54,000 hectares. In addition, both drainage and irrigation schemes are progressing along the Greater Zab, at Ishaqi, around the port of Basra, Shatt al-Arab, and Lake Hammar (to increase rice cultivation), with other smaller projects such as at Mosul. Nevertheless there are schools of thought that the greater pace of irrigation is too far ahead of the practical problems of drainage—one of the most serious of Iraq's difficulties in agriculture—and is leading to increased soil salinity.

Canals in olden times were constructed; a whole network, in fact, dates back to the Abbasid period (*circa* AD 850 to 1,000), with the object of watering arid areas and draining waterlogged ones by using the variation in the relative levels of the Tigris and the Euphrates. Many canals were built on both sides of the Tigris, and five large canals took water from the Euphrates to the Tigris, at that time to the area of Baghdad and Babylon. Old systems of canals have been opened up and cleared in addition to the new barrages and canals, but this revival did not take place until after 1880.

The Kut Barrage.

There has also been a vast increase in the attention paid to the problems of drainage, particularly within the last ten years. In order to lower the water-table in the subsoil, outfalls have been made which mean that ground-water can be reduced by being 'diverted' into rivers or land depressions, usually by means of pumps, though gravity flow is sometimes possible. Generally, dredging and some form of wind-breaks are needed to cope with the strong silt-carrying winds.

Obviously there are limits to the extent to which water can be diverted into depressions and allowed to evaporate. Nevertheless the problem has to be tackled of discharging drainage waters into the sea by constructing special canals and conduits.

Although the Iraqis still irrigate by the more primitive method of lifting water by *shadoufs* and *sakias*, and similar contraptions, today there is a marked increase in the use of mechanical pumps with diesel engines. More artesian wells, too, in the northern and southern deserts are bringing fertility to these arid regions. Obviously with the help of free-flow watering from barrages there has been a marked decline in the need for out-dated traditional water-lifting methods—about one-fifth now, as compared with 50 per cent twenty years ago.

A pipe inlet of subsidiary canals serving the land for crop irrigation by gravity flow, showing the gate closed.

268

REFERENCES

The Iraqi Journal of Agricultural Science, Vol. III, January 1968, No. 2, Al-Zahra Press, Baghdad

Michael Adams (Ed.), *The Middle East. A Handbook*, Anthony Blond, 1971

Abbas Ainsrawy, *Finance and Economic Development in Iraq*, Praegar, New York 1966

J. I. Clarke and W. B. Fisher, *Populations of the Middle East and North Africa. A Geographical Approach*, University of London Press, 1972

The Middle East and North Africa 1975-76. A Survey and Reference Book, 22nd edition, Europa Publications

More Water for Arid Lands. Promising Technologies and Research Opportunities, National Academy of Sciences, Washington 1974

International Bank for Reconstruction and Development, *Irrigation, Iraq*, Washington 1973

Marion Clawson, Hans H. Landsberg, Lyle T. Alexander, *The Agricultural Potential of the Middle East. Economic and Political Problems and Prospects*, Elsevier, New York, London, Amsterdam 1971

The Arabian Peninsula

The Arabian Peninsula includes the territories bordered by the Indian Ocean on the south, on the east by the Arabian Gulf and the Gulf of Oman, while on the west the Red Sea forms the third boundary of this Peninsula. The northern land boundaries are made up of the deserts of both Jordan and Iraq, and the whole Peninsula includes the countries of Kuwait, a wilderness of shifting sand which forms the Neutral Zone, Saudi Arabia, Bahrain, Qatar, the United Arab Emirates, Oman, the People's Democratic Republic of Yemen, and the Yemen Arab Republic, all of which are considered in greater detail under their respective chapter headings.

This remote peninsula is 2,000 kilometres long; its average width is 1,125 kilometres, giving it an area larger than that of India, or nearly two-and-a-half million square kilometres. Cushioned from the world until the discovery of oil transformed, for instance, Abu Dhabi into the richest country in the world, the simplicity and customs and traditions which have been formulated throughout the ages have not been superseded by the onset of wealth. The Arabs of the Peninsula are still as they have always been—themselves. Time is meaningless and the desert is always there ready to encroach and to claim back man's progress.

The Peninsula is characterised by the vast deserts which occupy the main portion of its area. Among these are the Ad Dahna, the Nafud, and the Southern Desert or Rub al-Khali. The soil in some parts of these deserts is fertile and makes good pastureland after rain has fallen. Other parts, however, especially in the Ad Dahna and the Nafud, are characterised by their shifting, wind-blown sands.

Arabia can be described as a plateau tilted down to the east. The highest mountains, both harsh and imposing, rise abruptly from the Tihama, increasing in altitude in the south, reaching more than 3,000 metres in the central Yemen Arab Republic before tapering off about mid-way in the area of Jeddah, Mecca and Medina. Thus it is easy to understand why these three centres were so situated, because geographically they form the most direct route across Arabia and the easiest inland way from the coast.

From west to east the enormous Arabian plateau slopes gently down until it reaches sea-level at the Arabian Gulf. Along the south-east coast in Muscat and Oman another mountain ridge, the Jebel al-Akhdar, some 2,980 metres high, forms a barrier between the region lying along the shore and the rest of the Peninsula.

The whole plateau of Arabia is corrugated with deep wadis caused by the erosive process of rivers. Many of these wadis are still covered with layers of clay deposits left behind by this fluvial action in the past. The largest sand-filled dry wadi is that of Hadhramaut, several hundred kilometres in length. Wadi Sirhan in the Nafud is another example.

There are no perennial rivers in the Arabian Peninsula. In the west of Arabia lava formations cause vast barren expanses of land called *harras*, which are typical of this part of the Arab World. Inhospitable in the extreme because of dry winds and aridity, the north of the Peninsula has an average rainfall of only 100 to 200 mm annually, the Yemen Arab Republic alone being favoured with good and reliable rainfall. Temperature extremes with maxima of 47°C in summer but frost in winter add to the general discomfort for man and beast. Because of these temperature variations, violent winds coupled with the unpleasant humidity of the coastal areas such as the Red Sea make life in the Peninsula very difficult.

Because of the tilt of the great mass of land eastwards, and the height of the mountain chains in the west, rainfall in the Red Sea area seems through seepage to percolate in the easterly direction, to reappear in the form of springs along the coast of the Arabian Gulf. This may be an indication that water, sorely needed in the general harsh environment of the Peninsula, is present underground. Already irrigation schemes to tap these possible supplies are under way, though results to date have been disappointing. Such a trial project has been carried out at al-Kharj in Nejd.

Thus life in the prevailing conditions is largely dependent on oases and wells, as the following pages will show.

The Kingdom of Saudi Arabia

The Kingdom of Saudi Arabia is bounded on the west by the Red Sea and on the east by the Arabian Gulf. The countries to the north are Jordan, Iraq and Kuwait, and to the south Oman and Northern and Southern Yemen.

Saudi Arabia, with approximately 2·2 million square kilometres of territory in Arabia, is part of the largest peninsula in the world (2·7 million square kilometres), and is like a continent surrounded by sea on three sides, and by a vast, inhospitable sand sea in the north. The physical environment dominates life in Saudi Arabia, but with the advance and exploitation of technology since the discovery of oil in 1933, even the desert—for centuries the domain of the nomadic Bedouin and his camel—is beginning to flower as modern scientific methods locate more water. In the absence of reliable statistics it has been conservatively estimated that the population, according to the United Nations figures, was 6·99 million in 1967. Environmental factors govern the distribution, particularly rainfall, underground water resources, and deserts. Thus roughly the population can be divided into the following: urban dwellers 24 per cent; nomadic tribes 21 per cent; settled agricultural workers 55 per cent. Unlike many of the developing countries in the Arab World, Saudi Arabia has a much lower urban concentration, existing towns tending to increase their population as a direct result of the oil revenues. Riyadh is the capital.

Saudi Arabia has been a tribal country since ancient times, not as a direct result of colonisation or any other aspect of rule, as has occurred in many other countries we have considered, but simply as an extension of nature's harsh geographical fostering of a way of life largely nomadic in the endless search for pastures and water.

Saudi Arabia, amid its arid plateau environment, is a centre or core for the aspirations and pilgrimage of millions in both Asia and Africa, the Mecca that existed before oil was even considered.

Today it is made up of four regions, each of which geographically could be termed a separate unit: the Nejd (the core of the peninsula); the Hijaz to the west, between Nejd and the Red Sea; Al-Hasa to the east on the Arabian Gulf, rich in oil; and lastly Asir to the south-west, extending as far as the Yemen.

Nejd

Nejd is the largest region and comprises the Central Arab Desert as well as many mountains, wadis and deserts in the central part of the country. Its land is not all barren, as was believed in the past, since from Huran in the north to the Euphrates there stretches a flat area called the Hamad which has good pastures and many springs.

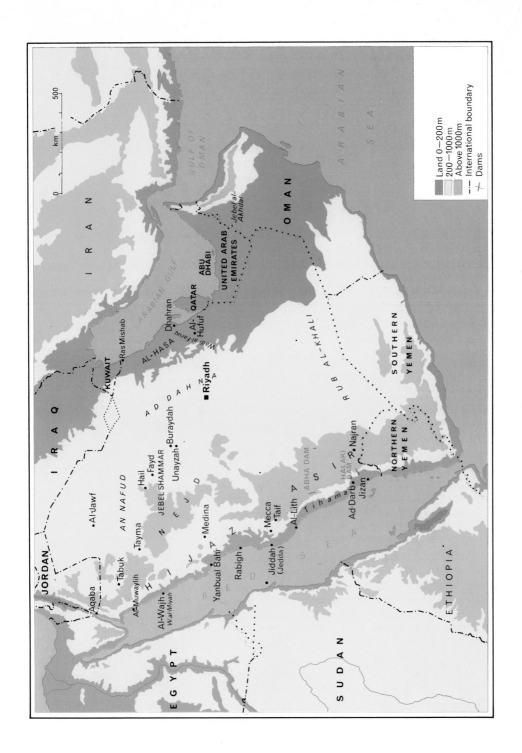

The Hijaz

This consists of a number of inter-related natural regions:

1. The coastal area stretching along the Red Sea (the Tihama) and fringed by coral reefs.

2. A high mountainous area which slopes gradually from Mecca down to the port of Jedda, where the altitude is below 300 metres.

3. An inland plateau area surrounded by mountains. In the north this plateau is very high and is covered with lava, but it slopes gradually down towards the south where, below Mecca and Taif, its altitude is only 300 metres.

4. The main ridge: the high parts of this ridge such as Kharma, Uwayrid and Khibar (1,800–2,000 metres high) are covered with lava. Near Mecca the ridge rises to a height of 1,500 metres.

5. The eastern slopes: these extend towards the centre of the Peninsula.

In the coastal and plateau regions are to be found a number of well-populated towns, like Al-Muwaylih, Wajh, Amlaj, Yanbual, Rabigh and Jedda in the coastal region, and Ula, Medina and Mecca on the plateau.

Many fertile oases are to be found in the Hijaz, situated mostly on a line between the main ridge and the eastern slopes; among these are al-Haiet, al-Huwait (Fadk), Khibar, Hanakia, Taif, Wadi Fatma and al-Safra.

Hasa

Hasa in the east was formerly known as Bahrain and Hajar. This name was applied to the area from Basra to Oman. Nowadays the name Hasa is given to the area stretching along the western coast of the Arabian Gulf, from Kuwait's southern borders as far as Saudi Arabia's frontiers with Qatar and Oman and the Khabura desert, where it is bounded in the west by the al-Summan desert.

Most of Hasa is a desert plain rising well above sea-level in the west, where the geographical features are similar to those of Tihama. There are a number of separate ranges of hills which form natural barriers between the regions. Eastwards, the plain gradually slopes down to the sea.

A range of hills extends towards the south along the Wadi el-Miah and the Jebel el-Taf, while the rocky al-Summan heights run parallel with the Arabian Gulf coast between Hasa and the sandy Dahna desert which separates this region from Nejd.

Wadi Faruq, a part of Wadi el-Miah in the south-west, is the most important wadi in this region. The coastal region is largely uncultivated but it contains many wells, in which the water-level almost reaches sea-level, and extensive pasture-lands; even the desert areas are modestly populated. Hasa and Qatif oases in the south are the richest areas in the region, and they contain numerous wells, springs and small lake-like river-basins.

Only the fertile part extending 20 kilometres east of Hufuf is under cultivation and is well-populated. In the north there are, however, some isolated areas of cultivation watered by springs. The Hasa region is rich in water resources and has many hot springs scattered over a wide area. Roads have been built near these springs.

Rice and other grain-crops are cultivated, but the principal crop is dates, of which there are several kinds. Other fruits grown extensively are citrus fruit, peaches, apricots, pomegranates, grapes and figs. Fine Arab horses, good breeds of donkeys and cows, as well as camels and sheep, are reared here in large numbers. The region is also rich in oilfields.

Asir

The name Asir is given to the south-western part of the Peninsula, south of the Hijaz mountains and north of the Yemen. This area coupled with the Najran region forms the best agricultural area in Saudi Arabia; most of Asir lies above 1,500 metres, reaching 3,000 metres at some points. The clue to the regions' fertility is that they are influenced by the monsoonal phenomenon of the Indian Ocean. Thus these uplands have a summer rainfall of 150 to 300 mm, which in addition to the elevation brings about dew and fog precipitation, all helpful to agriculture and giving a livelihood to many more people. Fast-running streams flow for several months of the year within deep valleys or *wadis* filled with a jungle-like bush vegetation in their upper reaches. This is gradually replaced by grassland and tall, swaying palm-trees called *daum*, and there is extensive terracing on the higher slopes, for the cultivation of such unlikely crops (when we consider the overall picture of Saudi Arabia) as yams, vines, wheat, millet, bananas, coffee and dates (the *daum* is an extremely hard variety). Where the uplands join the narrow coastal plain in the west, the streams disappear into the sand. Nevertheless this underground supply is put to good use in the cultivation of millet and dates.

Where the land drops towards the Red Sea, the pattern of aridity returns, with fringe land where nomadic herding of goats can be practised, but the coastland itself, the Tihama, can support but feeble scrub vegetation unfit for man or beast, and is desolate. This forms a great contrast to the fertile coastal plains we have considered in other countries.

Asir is the most beautiful part of Saudi Arabia, some of its lovely natural (almost tropical) scenery being similar to that of parts of Italy, Cyprus and the southern part of the Yemen.

GEOLOGICAL FORMATION

The country can be divided into two geological sections: namely, the Western Province (Hijaz) and the western part of the central province (Nejd), which consist of crystalline rocks; and the Eastern Province, which is mainly made up of sedimentary rocks, including mineral rock. These sedimentary rocks include, in particular, limestone. Sandstone, too, is found underground in Ula, in the northern part of Saudi Arabia. *Sabkhat* (salt flats) of the Eastern Province are typical, as are the *hadhabat* (gravel plains), and still more common are the areas of *dikakah*, which are hummocky growths in sand of bushes and grasses.

WATER RESOURCES

At present there is not much data on the potential water content in Saudi Arabia, indeed whether there is enough to provide for both present and future

In the neighbourhood of Asir, water for irrigation is drawn from the mountain torrents.

Rub al-Khali: the Empty Quarter. Too inhospitable in its wilderness of sand to support even nomadic life; monotony is relieved only by bare, weathered rock face.

requirements. Advanced technology has provided the means of producing sweet water from the sea at a high cost, but economically this is not a viable process for irrigation at the moment. Saudi Arabia has no rivers, only wells, springs, and rain-water courses. For the purposes of assessing the relative importance of water-resources, the country can be divided, according to climatic conditions and geological formation, into the following regions, all of which depend on rainfall and natural springs:

The South-Western Region
The Northern and Central Region
The region of springs, comprising Heet, Aflaj and Kharg
The Al-Hasa Region

Areas with natural springs are being developed and provided with irrigation and drainage facilities for cultivation, while the areas more dependent on rainfall are having their water regulated by the use of dams.

The South-Western Region

This stretches from the old town of Lith at the southern limit of the Hijaz, and includes the districts of Asir, Najran, Wadi Dawasir, Wadi Bisha and Sulaiyil. All these areas are irrigated from the torrents flowing down from the mountains. North of the city of Ad-Darb there are many wadis, from which the water irrigates the neighbouring areas.

The springs in Najran are rich in water but little use is made of them. The natural phenomenon of rocks and stones in Wadi Dawasir and Wadi Bisha indicates the possible existence of an ample supply of subterranean as well as surface water. In Sulaiyil, water is found near the surface.

The Northern and Central Region

This area includes the northern Hijaz, Nejd, and the southern parts of Al-Hasa.

The most important source of water in the Hijaz is Wadi Fatma, water from which is carried to Jedda by pipeline. Ein Zubaida, and recently Ein Hunayn, provide Mecca with water, while Taif is irrigated by wells and springs.

Two important wadis in the Hijaz are Wadi Haum and its tributary, Wadi Jazl. Water in both wadis can be found between 20 and 30 centimetres below the surface. The Khibar area has many springs which flow from among the rocks of volcanic lava that here extend deep underground. One of the important water-sources in upper Nejd is the Wadi Rama, which irrigates most of the orchards and palm-trees in Qasim, particularly in the Uniza area which extends along the wadi. The people of the area make their own drills for boring for the water from this wadi.

The Haiel area is irrigated by wells and springs and has rich pasturelands. The water in the wells is usually 10 to 25 metres below the surface, though in some parts it lies only 2 metres down.

The Region of Springs

The third region comprises Kharg, Aflaj and Heet. The earth's crust in this area,

The search for water is given priority.

Women drawing water from a well in the
desolation of an area given over to desert,
using methods that have survived for
generations.

though usually covered with sand, is formed of friable sedimentary rocks. As a result of pressures on the crust in ancient times, cracks have appeared into which water from streams on the slopes of the Tuwaiq mountains and the surrounding land has seeped, causing the formation of many large caves and water-holes. As a result a number of springs or *eins* are to be found, such as Ein Dala, Ein Khisa and Ein Samha, in Kharg; and Ein Raal, Ein Shugheib, Ein Botein and Ein Heet in Aflaj.

The Al-Hasa Region (see also p. 274)

One part of this region which is rich in water is the Birin Oasis, situated at the corner of the Rub al-Khali desert, known in the past as Ramlat Birin. The larger part of this oasis turns into a swamp in winter owing to lack of evaporation at that season, when the atmosphere is very much cooler and more humid than during the extremely hot dry summer.

AGRICULTURE

Agricultural projects are given high priority in Saudi Arabia because of the country's diverse problems, the great need for agricultural development and less reliance on oil, and the increase in urban population. The Government is careful to combine the budget for agriculture with that for water, and as a result it has increased enormously in ten years. FAO estimates show that 1·7 million hectares of land are forested while some 373,000 hectares are under cultivation of both arable and perennial crops. Dates are still the main crop. Climate is of tremendous importance, causing aridity throughout most of the land; less than 0·2 per cent is cultivated, and the potential according to Mallakh (1966) is probably about 15 per cent in total. High summer temperatures, as much as 47°C in the interior, added to a high degree of coastal humidity and an unpredictable winter rainfall, render cultivation difficult.

Farmers receive a great deal of assistance and encouragement from the Ministry of Agriculture and Water. This is evident from the establishment of many agricultural units in the various regions. Each unit has on its staff a number of technical advisers, is provided with stocks of machinery, improved seeds and fertilisers and also runs a model farm, an experimental and research centre and a nursery to produce improved varieties of seeds and saplings. For instance, ever since 1970, at the Animal Production and Agricultural Research Centre at Al-Hufuf, and at the Agricultural Research Farm at Deirab, intensive studies have been going on to improve livestock breeds and fisheries. The University College of North Wales has had a hand in these studies. The setting up of further research centres and experiments with, for instance, the more hardy Mexipak variety of wheat in collaboration with the FAO and the International Institute for Wheat Studies in Mexico have produced a yield more than double that of the traditionally grown wheat—no mean feat, wheat being the second largest crop grown. By the end of 1975, which was also the end of an agricultural five-year plan, an agricultural growth-rate of about 5 per cent a year was achieved.

Obviously there is a big demand for fertilisers because of the poor quality of the soil in most areas, and Saudi Arabia is fortunate in having sufficient for her own requirements. Research into soil deficiencies is being carried out. There is the distinct possibility that experiments in growing produce in artificially-controlled environments like those we have seen in Abu Dhabi may be emulated here as a method of agriculture that could flourish where otherwise external adverse conditions are 'anti-growth' of crops.

As an additional incentive the Government has also established the Arab Saudi Agricultural Bank in Riyadh with a capital of 30 million rials (£3 million); it has branches and regional offices throughout the country, and the farmers are encouraged to take advantage of the loan service provided. Of the loans made in 1972, 21·1 per cent were for buying machinery, 10·9 per cent for drilling, deepening and lining wells, 10·7 per cent for the purchase of pumps, and 9·5 per cent for new livestock.

The Ministry is also carrying out a number of projects, such as desert development in the north, aimed at helping the nomadic Bedouin, who suffer acutely from the scarcity of rain in the northern desert areas. The 1960–67 drought took a devastating toll of livestock. This deficit has now been made up by better breeds and by research into the use of dates as forage, for example.

The inhabitants of these regions are encouraged to cultivate the land using modern techniques, and thus to improve their conditions. To this end the Government offers farmers the following services:

> the loan of irrigation machinery and the provision of fuel and spare parts;
> the creation of cultivable land by the levelling and ploughing of plots of between 1 and 5 hectares around wells which the farmers have themselves dug;
> the distribution of seeds, fertilisers and pesticides for cultivating these areas with wheat, barley, clover and vegetables, and the training of farmers in modern methods of cultivation. The increase in the demand for vegetables has grown enormously. There is now an ever-growing market for water-melons, tomatoes, pumpkins, onions, egg-plants and other vegetables.

A monthly payment of 100 rials (about £10) is offered to every farmer who owns a private irrigation plant within the framework of the project.

The Faisal Bedouin Resettlement Scheme

In the central region between Riyadh and Hufuf, water on the surface is scarce and the Government found it advisable to exploit the results of the surveys conducted in the Hard area, which proved the existence of a large reservoir of subterranean water suitable for development as well as areas of potentially suitable arable land. Nomadism has always been basic to three million Bedouin, who form the backbone of animal husbandry in Saudi Arabia and whose allegiance is to the tribe. Thus the concept of a modern state is difficult for them to realise, with new values and ideas. Nevertheless the Government is encouraging them to become more settled and is teaching them new agricultural methods and techniques. The Faisal Bedouin Resettlement Scheme is but one such scheme on a larger scale which is meeting with success in rural settlement. This settlement is geared to sheep-breeding, with an initial flock of 7,200 Nejdi sheep being purchased and

1,000 hectares put under forage cultivation. The aim is to double the area culti-vated, and the flocks, in the next few years, with the provision of some 200,000 sheep a year (over four years) for local markets.

LAND TENURE

It is interesting to compare land holding with the other countries we have seen so far, as Saudi Arabia has a distinctively different policy. The proportion of agri-cultural owner-occupied holdings is somewhere in the region of 81·4 per cent in the eastern region, with an even higher percentage in the northern region, of 99·9 per cent. More than half the holdings are tiny, a mere 0·5 hectares, but there is obviously an entrenched independent ownership (with tribal allegiance), the size of holdings being of little importance until recently.

Distribution of Fallow Lands

A scheme for distributing uncultivated but cultivable land in Saudi Arabia was introduced in September 1968, broadly based on the following:

a. The area allotted to each individual for development had to be sufficient to establish an economically viable farm, i.e. a minimum of 12 and a maximum of 24 hectares. For companies and institutions it is 100 hectares.

b. Increasing the farmer's income and improving his living conditions, at the same time extending the cultivable area by land reclamation and by encouraging the investment of private capital in the land.

The inter-planting of citrus trees with other fruit such as figs is common practice where intensive use of fertile soil is possible, as in this orchard in Jedda.

c. An investment term of from three to four years; if the land has remained unworked or fallow during this period, it is liable to be allotted to another investor.

d. Distribution of land as soon as water and soil studies of an area are completed. Examples of areas where this type of scheme has already been implemented are: Kharg, Hasa, Haiel, Qatif, Bisha, Tabuk, Tihama, Ula and Qasim.

Measures against Sand Encroachment in Al-Hasa, and Irrigation and Drainage of the Region

Al-Hasa is a fertile oasis, rich in underground water with a low saline content, and in spite of its being surrounded by shifting sand on three sides, the south, the west, and the north, the al-Hasa agricultural scheme is the largest, and one of the most ambitious in terms of reclamation, of any region in Saudi Arabia.

Numerous springs and artesian wells provide water at two strata levels: the first 180 metres, the other 250 metres deep. The rise in the level of the underground water table has resulted in the flooding of some 8,000 hectares of land which were formerly cultivated and irrigated by primitive methods. The drainage of the excess surface water—typical, too, of the Qatif region where *sebkhas* have formed—so as to regulate irrigation in the area, will eventually reclaim some 20,000 hectares of land, cutting out some springs and wells with a lower water output and insufficient pressure in order to refurbish those with a greater output (the average is 1·7 cubic metres per second).

A complete network of irrigation and drainage canals, 1,623 kilometres of the former and 1,521 kilometres of the latter, serves the reclaimed land. Irrigation is by gravity, depending on the level of the land, or by pump. This scheme was inaugurated in 1971 and a maintenance organisation has been set up for the mechanical and civil engineering works in the area. The sand dunes to the north are a hazard because the north wind carries the small grains of sand left behind by the ebb and flow of the tide on the shallow shore of the Gulf into the Hasa oasis to the south-east. Thus large cultivated areas, and even entire villages, have been completely covered by sand, and many springs, canals and drains have been engulfed. It was here that the Ministry of Agriculture first undertook the study of this problem, a typical pattern where desert conditions are paramount, and a project is now under way which aims not only to protect the cultivated reclaimed land against wind-blown sand, draining the swamps formed by the filling-up of springs, but also to reclaim the areas already covered with sand, by planting trees and sowing grass to reduce encroachment and protect pastures.

The desert might be described as Saudi Arabia's most problematical enemy—it gets in the way of agriculturally productive land and sources of water, by splitting up these areas, dividing one from another by its vast empty wastes. The sand is continually shifting, engulfing attempts to halt its invasion into productive land and discouraging settlers. As in al-Hasa, trees are being cultivated in many areas, and grasses too, which when well-rooted act as a barrier against the creeping sand.

There are three main areas of sandy desert—the Great Nafud (5,720 square kilometres), the Rub al-Khali (780,000 square kilometres), and the Ad-Dahna (1,300 square kilometres), which connects the Great Nafud with the Rub al-Khali.

282

Dams

In areas where there is a dependence on rainfall, dams are being constructed to regulate the water for cultivation. And one of the most important projects is that of the Wadi Jizan. This dam lies at the extreme south-western point of Saudi Arabia, between high mountains culminating in a flat coastal plain running along the Red Sea, with a high degree of humidity. This site was chosen because use can be made of both the rainfall and the water from the mountain torrents. Average rainfall on the coast is 200 mm, 600 mm higher up. The dam rises 35 metres above the level of the water-course in the valley. This site would enable a reservoir to be formed above the dam with a capacity of some 51 million cubic metres of rain-water (rainfall during the summer season, from 1970 onwards, is already being used from this reservoir).

Prior to the construction of any permanent type of 'water containment', the farmers of the region used primitive methods of irrigation, with earthen or stone walls to try to divert the water to their particular land, in order to conserve it for the dry seasons and thus allow crops to be planted. However, the power of the water in spate generally swept away their efforts and, as we have already seen, the water was 'lost' in numerous streams before ever reaching the sea. Science has changed all this. The Wadi Jizan flood plain of about 7,000 hectares, coupled with the control of water for irrigation by the recently completed Malaki Dam and a smaller scheme relying on tube-well irrigation, will make it possible to grow a cash crop such as cotton in this area. Soil fertility is not a limiting factor, according to field studies which have been carried out, in areas which are irrigable. Research and development schemes for crop potential have gone hand in hand with this great project, and many of the villages are safe from the threat of floods.

Many smaller dams have been constructed such as the Mujamma and Dariya; three small dams have been built in the valleys branching off from Wadi Hanifa to maintain the well water, regulate the flow of mud sediment into the valleys, and provide enough water for a moist soil for agriculture. Other dams are intended to supply urban needs, like the one at Abha, and unique to Riyadh is its water tower which fans out in a dramatic design 61 metres above the ground, like a giant striped umbrella; it stands as a tribute to progress and twentieth-century technology in water storage.

International Co-operation

Unlike Kuwait and the Gulf States which rely on foreign labour, Saudi Arabia is neither an immigrant country nor an emigrant one; a pattern of self-reliance is practised, with the gaps in skilled labour being filled by the hiring of foreign expertise in the form of consultants or experts on a contractual basis, and with the additional help of the United Nations' specialised agencies. In this way it is hoped, with the passage of time, to build up a fully trained Saudi team who can take over the skilled work required, for example in the technological fields of irrigation and agriculture.

The Necessity for a Complete Land Survey

The discovery of oil along the eastern zone among sandstone and limestone strata has naturally provided the revenue to develop this vast country and to exploit its natural resources, but the lack of adequate maps has seriously hindered planners: development demanded geological exploration not only to find further oil but to find water for both urban and irrigation needs, to search for minerals, analyse soils, and to plan a communications network so urgently required. It is still some years before the mapping of the whole country will be completed by a photographic process. Even the vast sand seas of the Great Nafud and the Rub al-Khali are unfolding their secrets, and with them goes the somewhat mythical picture of soft, shifting sands (of which a very small part does exist), the geological truth revealing a panorama of soilless rocky mountains, boulders, gravel plains, and salt wastes.

The absence of water is not only an agricultural problem but is the limiting factor in the development of the country; oil revenues are being used to help overcome the harsh topographical conditions. Much of Saudi Arabia's future lies in the careful exploitation of what resources she has; the development of smaller-scale agricultural projects which can reach fruition may well prove to be the solution rather than larger projects which cannot meet with success in the limited cultivable areas which are scattered throughout the Kingdom. Agriculturally, Saudi Arabia's position is similar to that of Libya, both countries being in the fortunate position of possessing oil, the revenues from which can greatly assist in the progressive development of their natural resources and the creation of a better communications network.

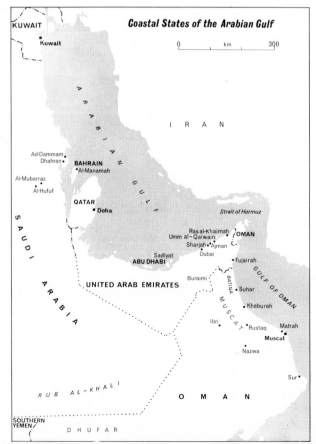

284

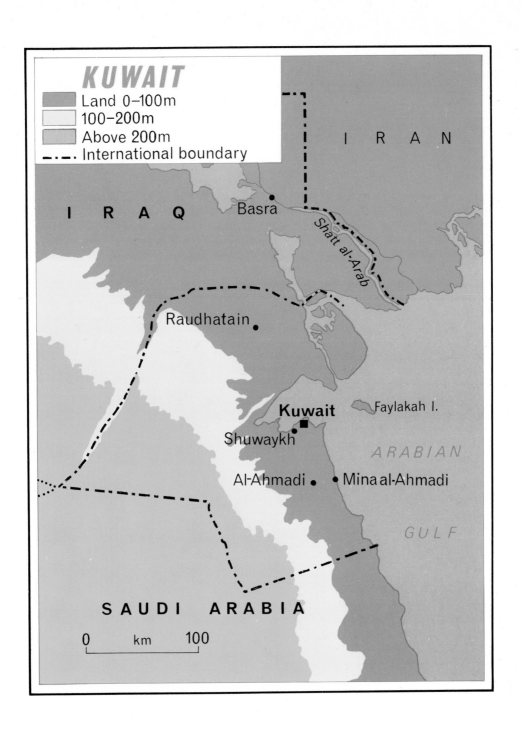

The Gulf States

THE STATE OF KUWAIT

The Emirate of Kuwait, one of the five Gulf States, compares with Bahrain, Qatar, the United Arab Emirates (formerly the Trucial States), and the Sultanate of Muscat and Oman in smallness of size and population. And in each case the rise of the oil industry dominates the way of life, sweeping away traditional methods which have been relegated to history in the wake of modern advances.

Kuwait is a roughly triangular country located at the north-western corner of the Arabian Gulf. It occupies an area of about 16,000 square kilometres, with an additional neutral or partitioned zone of 6,000 square kilometres which was intended originally for the passage of nomads and is administered jointly by Kuwait and Saudi Arabia, who share the mineral wealth of the zone. Kuwait's population in 1970 was estimated at about 800,000, which means that there are now almost three times as many people living in the country as there were before the discovery and exploitation of her enormous oil resources during the 1950s. This is a phenomenon associated with modern times: in growth it can be likened to Hong Kong, with an urbanised population increase that is remarkable; and the average wealth per head of population makes it one of the three richest countries in the world, with a high proportion of immigrants drawn to it (some 55 per cent) as to a magnet.

Kuwait consists of almost flat desert with an occasional elevation like the Ahmadi Ridge, which is about 120 metres high. Historically, the country has played a part in acting as a kind of bridge linking southern Iraq to the heart of the Gulf area. There are a number of small off-shore islands to the north and east, notably Faylakah, which has proved to be a source of archaeological interest. Average annual rainfall ranges from 25 mm to 175 mm, practically all of which falls between October and April. Summer temperatures can rise well above 50°C, but during the winter months temperatures by day range from 10°C to 15°C, and from time to time night frosts may occur.

Before the discovery and production of oil, the city and harbour of Kuwait had grown up as a centre for pearling and fishing and as a small market for trade between the nomadic populations of the hinterland, the coastal fishermen and merchants bringing in food and other goods in exchange for Kuwait's own limited range of products. Until very recent times, there was no way in which Kuwait, with her apparently scanty natural resources, could move forward or significantly develop, and no one thirty years ago could have foreseen that this country would in the 1970s have become one of the major oil-producing nations. Oil is the very life-blood of modern Kuwait and has transformed her almost overnight into one

286

of the richest nations in the world, with her wealth being used not only in the creation of a modern welfare state but in helping her sister Arab countries through the Kuwait Fund for Arab Economic Development.

But Kuwait, in common with all the dry undulating desert countries of Arabia, has always had a central problem of the lack of water. There are no rivers in Kuwait, no lakes, no sources of water except a few small oases and springs. Such settlements as there were before the discovery of oil grew up where underground water lay and aquifers made small-scale agriculture possible. Indeed, in Kuwait generally, the water shortage was so acute that for years water had actually been imported in *dhows* from the Shatt al-Arab in Iraq. After being landed near the present port of Shuwaykh, it would be sold to carriers for distribution, but the whole operation was very costly and the purity of the imported water often suspect.

With the rapid development of the oil industry, the Government has made the provision of ample supplies of both drinking-water and water for irrigation one of its first priorities. Drinking-water is now piped direct from the Shatt al-Arab; and by using waste gas, a sea-water distillation plant at Mina al-Ahmadi is able to produce some 80 million litres of drinking-water daily, much of which is piped across the desert to Kuwait City. The Government, following an oil company venture in 1950, then installed its own plant at Shuwaykh, producing 4 million litres of water daily, this plant being opened in 1953. Since then, further government desalination plants have been built, and output of water is likely to amount to the staggering total of 120 million litres per day. In addition, important new sources of fresh water have been found by tapping the underground water-table at Raudhatain and al-Shiqaya.

Of course, the greater part of all this 'new' water is used for domestic consumption and for the rapidly-growing industrial development of Kuwait; but agriculture is not neglected. Land control laws ensure that land remains in the hands of the government and the Kuwaitis, with non-Kuwaitis allowed only tenancy rights. About 3 per cent of the total land area is suitable for agriculture, but on the whole agriculture in Kuwait is not commercially viable. However, with the increase in water supplies by desalination and fertiliser plants in operation, and with experimental farming methods going on, the Government hope to achieve results in obtaining crops and saplings (for halting erosion) which are sturdy and able to withstand soil salinity and the type of brackish water prevalent. Improved breeds of livestock are being reared not only to assist the Bedouin to improve their camel, sheep and goat stock, but to encourage the tiny rural community, in addition, to increase dairy farming and poultry rearing by the battery system. The fishing industry still lives on in spite of the heavy industrialisation of such an oil-rich country, and fish processing plant is but a natural consequence of the progress made since the discovery of oil.

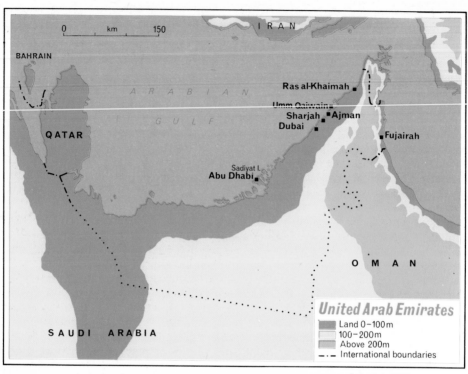

This desalination plant in the U.A.E. not only provides electricity but enables agriculture to be practised where water was previously unobtainable.

Onion vendor in Dubai.

Date plantation in Bahrain.

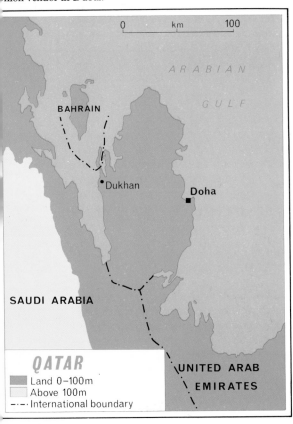

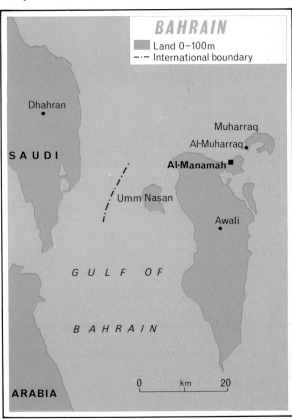

THE STATE OF BAHRAIN

The Emirate of Bahrain, which became independent in 1971, consists of an archipelago of thirty-three small low-lying limestone islands off the east coast of Saudi Arabia and the Qatar Peninsula. The total area of the group is approximately 662 square kilometres (little larger than the Isle of Man) with a population of some 218,000. The principal island, Bahrain itself, which contains the capital city of Manamah, is 48 kilometres long and 16 kilometres wide, while other important islands are Muharraq, Sitra, Nabih Salih, Jiddah, Huwar and Umm an-Nassan. Many of the remaining islands are either uninhabited or contain only tiny human settlements.

In spite of being so tiny, Bahrain was one of the first Gulf states where oil was discovered. This was in 1932, and the state now boasts the second largest refinery in the Middle East. The oil industry, coupled with its rapidly increasing importance as a trading centre and entrepôt, is the chief mainstay of Bahrain's economy today; but nonetheless agriculture and animal husbandry are widely practised throughout the islands, which contain numerous and prolific sweet-water springs and natural artesian water supplies, particularly in the northern part of the archipelago. These make possible widespread cultivation of dates, vegetables, and fodder-crops for cattle-breeding. It is here that villages have clustered and developed their agriculture.

Before oil was developed life was hard; nomadic herdsmen found some pasture-land, but cultivation was much more limited and confined to the oasis type. Small, scattered communities relied for their living on areas where there was more water, while fishing and pearl-diving provided a different form of occupation.

The Government nowadays runs an Agricultural Experimental Centre and model farm which provides farmers with numerous services and advice: for example, the results of experiments with new types of crop and equipment for up-to-date methods of agriculture and pest-control. The more efficient use of ground-water is also encouraged in order to assist the farmers, most of whom are Bahraini (50 per cent of the population now are non-Bahraini).

Rainfall in Bahrain is slight, averaging under 75 mm a year, so that were it not for the ample spring and underground water with which Bahrain is blessed, its agricultural activities would be far more restricted than they are, and irrigation would be necessary on a much wider scale. But this is not needed, and Bahrain has not yet had to face the problems of acute water shortage that have afflicted the desert nations of the Arab World.

THE STATE OF QATAR

The Shaikhdom of Qatar, which became completely independent in 1971, consists of a large narrow peninsula projecting north from the west coast of the Arabian Gulf, dominated by the Jebel Dukhan, 70 metres high. It is approximately 160 kilometres long and varies between 50 and 80 kilometres in breadth. Its total area, including a few small off-shore islands, is about 22,000 square kilometres, although its frontiers with Saudi Arabia and Abu Dhabi have never been clearly defined. The population was estimated in 1970 at 130,000; results of the recent census of population have yet to be published, so today's total is not yet known.

Pruning guava trees in Qatar. The increase in citrus cultivation has been quite dramatic despite high mineral content of such water as there is. Desalinated water is piped for irrigation.

Poultry farming in Qatar. Part of the acceleration in the drive for self-sufficiency.

As we have already seen with, for example, Libya, the exploitation of oil has radically altered the economy of the country. The familiar pattern of urban growth is seen particularly around the capital city of Doha, on the east coast, where three-quarters of the population congregate. In a country which was made up mostly of semi-nomads and fishermen, oil has contributed its revenue towards improved agriculture of the market-garden type which relies on irrigation. This type of farming can be found particularly north-west of the capital city.

The territory of Qatar consists almost entirely of rocky barren desert, flat except for some low hills along the west coast and a central limestone plateau. In the south, there is little but sand and salt flats; these salt flats or marshes (*sebkhas*) are caused mainly by evaporation of water which may seep to the surface from underground or be fed by the sea. The end result is the same: marshland with a surface encrusted with salt, a barrier certainly not only for man but for agricultural potential. Such natural vegetation as Qatar possesses is confined to parts of the north after the winter rains, although the average rainfall is slight. The summer temperature is continuously very high from May to September, though from December to March the temperature is relatively cool, varying from 25°C to 33°C, but humidity is high throughout the year. There are very limited supplies of underground water (the only natural supply of water is obtained from wells), and many of these are unsuitable either for drinking or for agricultural purposes because of their very high mineral content.

Here again, as with Kuwait and Abu Dhabi, the potential for future development was very small until the discovery of oil in 1939, the exploitation of which has enabled Qatar to enjoy a good revenue (£75,000,000 in 1971). As a result, the Government is able to finance an ambitious development programme, including agricultural research and development, and to diversify the economy as far as possible. Because of the very nature of the country, this diversification is problematical. The problem of water-supply is being overcome to some extent by sea-water distillation plants, and Doha now has a piped system.

The Qatar Department of Agriculture has achieved startling successes in the past decade in the production of fruit, in afforestation, and especially in the intensive cultivation of vegetables. The area of pasturage and land under cultivation is about 139 square kilometres, according to the Food and Agriculture Organisation's figures. Qatar is now self-sufficient in vegetables, and there is even a surplus for export to other Gulf States. None of this, of course, would have been possible without adequate water supplies, and these in turn have only been forthcoming as a result of projects such as the highly expensive desalination process, which could not have been embarked upon without the vast expenditure made possible by the discovery of oil.

The need for fertilisers (but one example of modern trends) was also recognised by the Qataris, and urea and ammonia fertiliser plants are already in operation around the capital, as is a shrimp processing plant (part of the diversification policy). And the surveying for additional water continues as part of the scheme to produce more food and to help feed the rapidly increasing population as well as to provide educational and training facilities for the future.

In Abu Dhabi, experiments are being successfully carried out in cultivation under extreme arid conditions. Sadiyat Island has its own research centre.

Modern technology using materials such as polythene has assisted in transforming the desert in Abu Dhabi.

Maize under irrigation. Banks of earth trap the water and channel it to where it is required. Trees act as barriers against erosion.

THE UNITED ARAB EMIRATES

The United Arab Emirates, constituted as an independent federal state in December 1971, embraces the seven emirates of Abu Dhabi, Dubai, Sharjah, Ras al-Khaimah, Umm al-Qaiwain, Ajman and Fujairah. The total combined area is estimated at about 84,000 square kilometres, and the population in 1968 when the first census was held was 180,184. Its coastline extends for almost 650 kilometres, from the frontiers of Oman in the east to Qatar in the west, and into the sand desert of the Empty Quarter, with large areas of *sebkha* which lie along much of the coastal regions of Abu Dhabi. The climate is arid in the extreme. Summer temperatures are very high, but the winter temperature, ranging from 18°C to 24°C, coupled with the absence of the dust-storms which are frequent in the northern Gulf, enables the United Arab Emirates to enjoy a climate in winter which is as pleasant as that of some of the established tourist areas of the world, such as Nairobi or Miami.

Most of the population live in the oil-rich areas of Abu Dhabi and Dubai, since here again the whole life of the region has been drastically transformed by the discovery of oil. Abu Dhabi, which can claim to be one of the richest countries in the world in terms of income per head, at present contributes the lion's share of the revenues of the States (in 1972 it was 82 per cent) with most of the rest coming from Dubai, which has the role of being the major commercial port for all seven states and where off-shore oil has been discovered. Thus Dubai shares Abu Dhabi's prosperity.

Agriculture in the United Arab Emirates is largely confined, so far, to a few desert oases in the hinterland, with Ras al-Khaimah the main 'market garden' of the Federation. In pre-Federation days, the Emirate of Ras al-Khaimah pioneered research and experimentation into methods of growing fruit, vegetables and fodder crops. The cultivation of these was easier because the average annual rainfall is about 150 mm, higher than elsewhere in the Federation, while there is good arable land along the coast and among the foothills of the mountains. A great variety of crops are now cultivated: many kinds of fruit and vegetables, as well as animal fodder and some tobacco for local consumption. The Agricultural Trials Station at Digdagga in Ras al-Khaimah, established in 1955, is engaged in continuous research into methods of cultivation and the raising of livestock; and it possesses the only agricultural school in the United Arab Emirates, where students from all parts of the Federation can take a three-year practical and theoretical course.

The bulk of the population by 1968 (about 55 per cent) had become urbanised, living mainly in the towns of Dubai, Sharjah, and Abu Dhabi itself. The rest were mainly concentrated within the two coasts, with the exception of the Buraimi oasis (see Oman), where a large concourse of people have settled in the town of Al-Ain.

Abu Dhabi, which is by far the largest as well as the richest of the emirates, consists mainly of desert and scrub, but its hinterland is valuable as it contains oases such as Buraimi, with fresh-water supplies from the nearby mountain range. Abu Dhabi Town itself is one of the two hundred islands in this region, many of which are uninhabitable. Great progress has been made in Abu Dhabi since the establishment of an experimental farm at Al-Ain in 1968. Forestry, too, has been receiving attention and nearly 1,000 hectares of trees have already been planted, with further development envisaged in a bid to reclaim the desert. At Al-Ain, as is so often the case in the Land of the Arabs, the limiting factor is the supply of irrigation water, but it is estimated that there is enough water available to irrigate an area of over 3,000 hectares.

The Department of Agriculture is also paying much attention to educating farmers in new methods of cultivation and is actively encouraging the establishment of new farms. Local nationals are being given three-quarters of a hectare of land free of charge, together with all the necessary equipment and maintenance grants to tide them over the early period before their farms can become profitable. About 600 such farms have been established so far. Here, too, efforts are being made to persuade the pastoral Bedouin to change their nomadic way of life to that of settled farmers, but inevitably this is a slow process. Tribal housing units have been formed with the idea of keeping people together until they are more settled.

Elsewhere in the United Arab Emirates, with the exception of Fujairah, where a narrow strip of fertile land runs along the Batina coast, agricultural activity is relatively unimportant. This is obviously because of the nature of the country, and because most of the ancient wells dug by nomads contain brackish water. There are a few sweet-water wells, however, and these are being restored where they have fallen into neglect, and new ones are being dug, many of them equipped with diesel pumps. Apart from wells there are also numerous underground canals, known as aflaj, which tap the water-sources in the hills and lead the water down through tunnels to farms often many miles away. Many of these aflaj are very old, yet they were so well and ingeniously constructed that some of them have been cleared and renovated for use again. In complete contrast to ancient wells and aflaj, it is here that one of the most modern techniques of arid-zone technology is being used. This is at the Arid Lands Research Centre on Sadiyat Island off Abu Dhabi, which consists of an integrated complex for the generation of electricity, for the desalination of sea-water and for food production. The fresh water obtained from this complex is used to irrigate vegetables planted inside greenhouses made of plastic material. Both the amount of water used and the temperature inside the greenhouses are closely controlled and varied according to the nature of the crop and for research purposes.

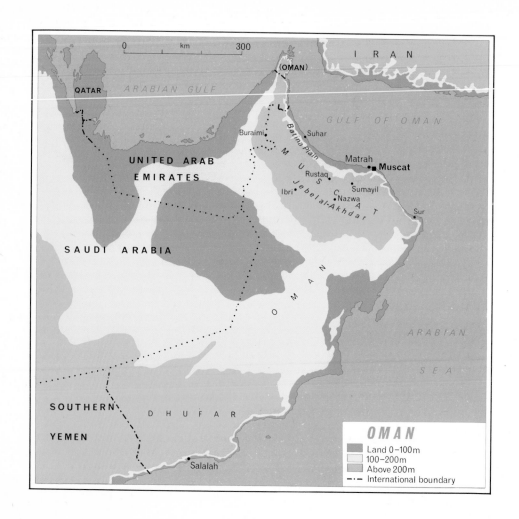

The Sultanate of Oman

The State (or Sultanate) of Muscat and Oman lies at the south-east corner of the Arabian Peninsula, and is by far the largest of the Gulf States. It is bordered by the United Arab Emirates and Saudi Arabia to the north and west, and by Southern Yemen in the south-west. Its coastline lies on the Gulf of Oman and the Arabian Sea. Its area is about 212,400 square kilometres (the landward boundaries are not defined). Data for the mixed racial population in 1969 vary between 500,000 and 700,000 people, with equally limited statistics available for their distribution. Muscat is the capital and chief port. Oman has a coastline extending for over 1,000 miles, and fishing has always been and continues to be important to its economy.

Oman was for centuries a country of mystery, its existence bonded to both the desert and the sea; its difficulty of access by land and the rugged, broken nature of its mountain ranges, particularly the Jebel al-Akhdar range in the north-east, acted as natural deterrents.

The mean annual rainfall at Muscat is just under 100 mm, and the average mean temperature ranges from 20°C to 35°C. Over most of the country water resources are not abundant, though they have not been fully assessed. Nevertheless, there are areas of relative fertility, notably the coastal plain of Batina and the plateau of Dhofar to the south-east. The coastal lowlands, mainly the Batina, have an approximate rainfall of 150 mm annually, with much more occurring in the highlands, which consist of craggy, sharp limestone, heavily dissected and soilless in a sombre landscape; yet there are, by way of contrast, other areas of green fields and groves of trees. Springs from the hills give the Batina scope for irrigation from streams or by *aflaj*, so that the villagers are able to cultivate a variety of commodities, mainly date and coconut palms, with banana, citrus fruits, pomegranates, cereals and vegetables to add to the favourable marine products like sardines and crayfish. With the help of irrigation and terracing of hillsides on inland valley floors, dates, figs and coffee can be grown, while limited pastures are available for semi-nomads at higher altitudes. It is provisionally estimated that about 40,000 hectares, or about 15 per cent of the land area of the country, are under settled cultivation, most

A typical wadi in Oman. The rainfall being so limited and unpredictable, agriculture depends on irrigation: in the interior, oases rely on a system of underground water channels known as *aflaj* which tap the water table above cultivation level; on the Batina coast wells provide water which is pumped up mechanically.

A saline quicksand in Oman.

of which lies in the Batina Plain. In the upland cultivated areas wheat takes the place of bananas and tobacco, while in Dhofar coconuts are also grown, and cattle-breeding is extensively practised. The Omani camel, which is bred in all parts of the country, is much prized throughout the Arabian Peninsula.

Prior to the discovery of oil, most of the Omani population could be categorised as either oasis-dwellers or nomads earning their livelihood from the land or the sea. Nowadays, the chief fact about the modern Omani economy is, as in the Gulf States, the production of oil. It was a late starter in the oil-producers' boom, not generating any oil until 1967, and up to that time retained its traditional identity, which was assisted by its rather more isolated location and its lack of enthusiasm for external influences. Although Oman's oil reserves and production are modest compared with those of other States, they are sufficient to ensure a moderate development programme. The Government is seeking to achieve this by investing in projects which will encourage, *inter alia*, agricultural development.

One of the most urgent problems facing Oman is, not surprisingly, its shortage of water, but, if all available supplies can be fully exploited and harnessed, it should be possible greatly to extend agriculture throughout the country. The Department of Agriculture has drawn up a five-year development plan, and high on the list of priorities is a comprehensive water-resource and hydrological survey, which is being conducted simultaneously with a land-use and classification survey. Conservative estimates indicate that a total of 100,000 hectares could be brought under cultivation by full and up-to-date use of Oman's water-resources. The five-year plan also encompasses research into new varieties of crops, training farmers in new methods, research into marketing of products, and an agricultural research institute.

Oman is certainly a land of contrasts. Seasonal streams, unexploited ground-water, and oases are potentially developing areas for agriculture. The existence of water is the dominant factor, as we have already seen in so many other developing countries in the Arab World intent not only on survival but on the expansion of food resources. Not only in agricultural terms, but in all fields of development, Oman may perhaps be said to be changing at this time more rapidly and radically than any other state in the Arab World—possibly in the whole world.

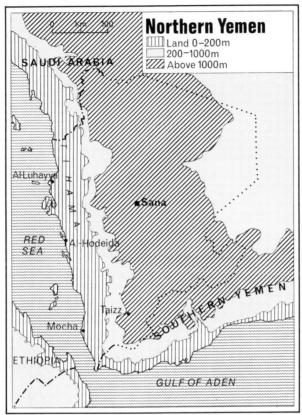

Northern Yemen

Land 0–200m
200–1000m
Above 1000m

SAUDI ARABIA

TIHAMA

AlLuhayya

Sana

RED SEA

Al-Hodeida

SOUTHERN YEMEN

Taizz

Mocha

ETHIOPIA

GULF OF ADEN

The Arab Republic of Yemen
(Northern Yemen)

The Arab Republic of Yemen, independent since 1918 when the Turks left, lies in the south-west corner of the Arabian Peninsula. As yet there has not been any census of population, so that a figure of five million, as given in the *United Nations Demographic Yearbook* for 1968, would be approximately correct for today, though it was but an estimate even at the time of publication. (Since the end of the Ottoman rule, outside visitors to the country have been few.) The Yemen does not have exactly defined frontiers with Saudi Arabia, so that to quote the area as 195,000 square kilometres is but an approximation also.

It falls into two distinct areas: the Tihama coastal strip, 25 to 40 kilometres wide, along the Red Sea; and the inland highland area. The capital Sanaa lies in the highland plateau, where there is widespread cultivation in the valleys of this loess-covered basin. Obviously the population tends to concentrate in the more fertile areas. The coastal regions are hot and humid, with under 120 mm of rainfall per

Climate plays a large part in the agricultural development of Northern Yemen, with perennial rivers flowing in the highlands but disappearing before the coast is reached. Crops are grown on the fertile alluvial soil and modern irrigation methods here form an interwoven pattern of water-carriers and reservoirs.

Some of the most fertile land in the Arabian Peninsula, in Northern Yemen. Extensive terracing prevents soil erosion, and in the area of Sana crops such as coffee, *qat*, cereals and vegetables are grown.

The Wadi Hadramaut in Southern Yemen. In the upper reaches, where the valley is broad and flood water is available after rain, cultivation is possible.

year, and little vegetation is found there apart from the date-palm. They are generally sandy with some *sebkha* and fertile silts and clays from the highlands. High temperatures and searing winds can make cultivation difficult. Cotton is being developed here as in other areas, and is exported. But it is really above the foothills at around 750 metres that conditions for cultivation are at their most profitable, because of the seasonal flow of streams, coupled with strata seepage, which make the regulation of water possible. Where the land is not flat, extensive terracing has been carried out, so that a wide variety of crops can be grown in a zone which avoids extremes of temperature, and it has been developed on the sides of many of the wadis.

The highlands enjoy probably the best climate in all Arabia. Rainfall in the highest mountain areas may reach as much as 890 mm a year, while the annual average on the eastern high plateau is from 380 mm to 500 mm. These fertile highlands provide the Arab Republic of Yemen with the basis of its predominantly agricultural economy. It is this area which was known in ancient times as 'Arabia Felix', simply because of its great fertility, and it is thought to be in this region that the realm of Sheba was situated.

Although there is enough rainfall in the hill areas for dry farming crops like millet and sorghum, which are the chief traditional ones grown, around Sanaa and Taizz vegetables such as tomatoes, cauliflowers, lettuce, peas, and potatoes flourish; and wheat, barley and maize are also grown on the lower slopes of the mountains. Fruit-growing is widespread, with grapes, oranges, bananas, lemons and pomegranates flourishing in the highlands as well as dates along the coastal plain of Tihama. The most important export crops are coffee and cotton: the Mocha district produces almost half a million pounds' worth of coffee each year (the Yemen is the only country in the world which produces coffee on land which is irrigated— the famous Mocha coffee), while 12,000 tons of cotton were exported in 1971. *Qat*, a narcotic shrub, is very widely cultivated, although in 1972 the Government brought in measures to limit its cultivation and, indeed, to ban it on Government or *Wakf*-owned land. The *qat* shrub is unique to Northern Yemen and Southern Yemen. It grows nowhere else, and when the leaves are chewed acts as a mild narcotic (though its leaves retain these properties for only three days). This can perhaps be likened to the habit of tobacco-smoking or the consumption of alcohol in some other countries.

One might imagine from this brief description of the Northern Yemen that from the point of view of agriculture, at any rate, the country truly merited the epithet 'felix'. But there are problems. Despite the large area over which cereals are grown, the yield is low and staple foods still have to be imported. The rainfall, though generally abundant compared with the rest of Arabia, is not entirely reliable, and drought occurs frequently. There was a prolonged period from 1968–70 which caused actual famine and forced the Government to import very large quantities of cereals. A feature is the shifting cultivation which is carried out in some areas. When the rains fail in one valley, the population shifts to another area where rainfall is more bountiful, helping out with cropping and returning home with a percentage of the profit. About four-fifths of the cultivated land is rain-fed, while the remainder is irrigated (see diagram). Unfortunately, oil has not been discovered in the Northern Yemen, which makes her situation economically less prosperous than that of other Arabian States, so that she must rely very much on agriculture and on her developing, but still relatively small-scale, industries. Some Yemenis are seasonal workers, travelling to the Sudan, for instance, to pick cotton, or to Iraq for date harvesting.

It is therefore a major concern of the Government, aided by the United Nations and the Kuwait Fund for Arab Economic Development, to push forward projects designed to increase facilities for water-storage, to introduce more efficient methods of irrigation, and to make use of the underground water which is known to exist in the Tihama Plain. Specific projects currently under way are a full survey of the Wadi Zabid area, a pilot scheme for developing a sugar-cane industry, the establishment near Taizz of an agricultural advisory centre, and detailed schemes for developing and modernising farming both in the highlands and in the plain.

While it could be said that most Yemenis are engaged in farming at almost subsistence level, the wind of change due to governmental action is being felt. Prior to this the Imam and a few others held large areas of land as landlords, leasing them to tenant farmers with complicated crop-sharing procedures according to the type of crop the tenant farmed, the landlord in some cases contributing seed, machinery, and water in return for rental and a crop percentage. However, since 1962 the scene has changed somewhat, with state enterprises being set up and fixed wages being paid to workers, while the power of the landlords has lessened. These are quite distinct from the small communities of peasant farmers, and make a contrasting pattern to the animal husbandry found both within agricultural settlements and without, where nomadic herdsmen still practise as they have throughout the ages.

Thus with its relatively high population when compared with Saudi Arabia, the Northern Yemen is distinctly different from many of the countries which we have already looked at: there are no large cities and no particular drift towards urbanisation. Between 85 and 90 per cent of the population are settled rural dwellers, with approximately 5 per cent nomadic herders, and the remainder living in the few towns. Nevertheless, with the increase of development schemes and the expansion of ports like Hodeida, perhaps there will be in the future a more distinct trend towards urbanisation.

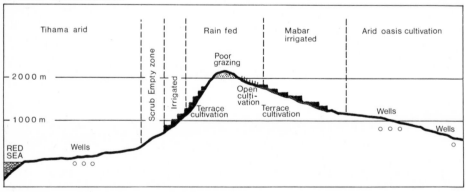

Northern Yemen has a unique pattern of land use.

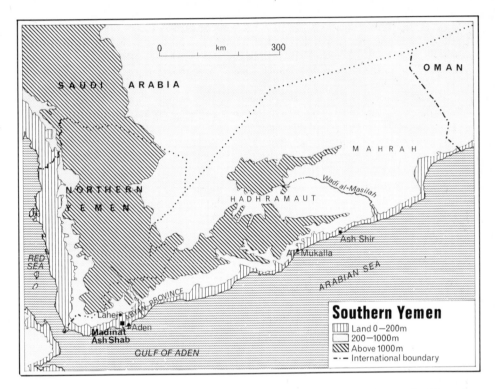

The map shows labels: OMAN, SAUDI ARABIA, NORTHERN YEMEN, MAHRAH, HADRAMAUT, Wadi al-Masilah, Ash Shir, Al Mukalla, ARABIAN SEA, RED SEA, ADEN PROVINCE, Lahej, Aden, Madinat Ash Shab, GULF OF ADEN.

Scale: 0 — km — 300

Southern Yemen
- Land 0—200m
- 200—1000m
- Above 1000m
- ·—· International boundary

The People's Democratic Republic of Yemen (Southern Yemen)

Southern Yemen stretches along the south-western coast of the Arabian Peninsula. It is bounded on the north by the Arab Republic of Yemen and the Kingdom of Saudi Arabia, to the east by the Sultanate of Oman, and to the south and west by the Gulf of Aden and the Arabian Sea. Its frontiers with Northern Yemen and Saudi Arabia have still to be settled, so its total land area is about 290,000 square kilometres, with an estimated population of about 1,500,000. The Republic, which was formed in 1967, also owns several islands such as Socotra, near the Arabian coast. Its capital city is the famous port of Aden, which has for centuries been of enormous strategic and military importance. The closure of the Suez Canal affected Aden City disastrously from the economic point of view, as did the consequent development of super-tankers which could use the Cape of Good Hope route. A refinery, used for the import and export of oil, made the most important contribution to the trade of Southern Yemen. Refining still continues. Even with the reopening of the Canal, the future does not look very certain.

The greater part of Southern Yemen consists of desolate plains and mountains and is practically devoid of vegetation. Cultivation is therefore confined to scattered and relatively small patches of fertile soil in some of the river-valleys, especially in the Wadi Hadhramaut, which runs parallel with the coast about 160 kilometres inland, and has partial expanses of level plain, where irregular rainfall causes flooding (*seil*) and gives rise eventually to surface water. The upper part of this great wadi is under settled cultivation owing to the alluvial soil deposited by the *seil* flood-water. Elsewhere there is land made cultivable by traditional methods of irrigation from wells and from cisterns hollowed out of the rock; some of these are of great antiquity.

The climate of the country is very varied. In Aden itself, about 127 mm of rain fall each year in the winter months; in the highland areas that stretch north of Aden eastwards to the province of Dhofar, annual rainfall varying from 380 mm to 760 mm occurs, mostly in summer; while in the extreme eastern lowlands and in the northern deserts rainfall is so rare and scanty as to be for practical purposes almost negligible. Temperatures in the country are universally high, with extreme humidity in the south-west coastal zone, and dry desert conditions in the far north.

Only about a quarter of the potentially cultivable land of Southern Yemen is at present used, and one of the main aims of the Government's Three-Year Plan recently completed was the expansion of agricultural production. Most of the population are engaged in agriculture in one form or another, and most of the crops are of a subsistence nature. Wheat is grown in the Hadhramaut, and other cereals include barley, millet and sorghum. In most vegetable and fruit crops the country is largely self-supporting, while the Food and Agriculture Organisation has recommended an increase in banana-growing as a potentially valuable export. Fishing forms a secondary industry, Mukalla and Ash-Shihr acting as fishing centres. Vast quantities of tunny, sardine, and sharks are caught, many of which are dried in the sun for their oil.

Cotton is grown mainly in the intensively cultivated areas of Abyan and Lahej, to the east and north of Aden, and the most flourishing agricultural co-operatives in Southern Yemen are those of the cotton-producers. Raw cotton is one of the major exports in spite of marketing problems and increasing competition. Lahej is also responsible for cereals, dates, bananas, coconuts, vegetables, coffee and *qat*. But on the whole, cultivation, because of lack of good soil and water, is much more scattered and less developed than in the Northern Yemen.

The improvement of the varieties of cotton produced is going hand in hand with pest control and mechanisation. Tobacco is grown in coastal areas, and frankincense trees in their natural state flourish above 1,000 metres on the southern plateaux, where the rainfall is slightly heavier. The sap is collected for the production of myrrh and incense, generally before the summer rains start. Livestock production, chiefly of sheep and goats, is being encouraged.

Seil cultivation has always been a traditional method of irrigation in the Southern Yemen. It operates rather as the basic type of irrigation which Egypt used before the 'harnessing' of the River Nile. Low dams or barrages are built which divert seasonal flood-water on to selected areas of ground suitable for cultivation. The areas chosen vary in rotation so that maximum benefit can be obtained from the use of such water as there is; slightly less reliance need therefore be placed on the erratic rainfall. *Seil* cultivation requires somewhat narrow valleys for its effective use. It is also practised in the Northern Yemen and eastern Sudan, and is not confined to the Wadi Hadhramaut. The Abyan cotton irrigation scheme is a good example of the effective use of this type of crop rotation.

Overall, in the field of agriculture, the Government aims to increase the cultivated area of the country by about 8 per cent per year and to expand production of crops and livestock by some 20 per cent. There is not enough finance available to put development projects into effect in all the possible locations throughout the country, and current efforts are concentrated in those areas that are already centres of cultivation. The successful expansion of agriculture will obviously depend very much on the introduction of modern irrigation methods and machinery, and this is among the Government's highest priorities.

Nevertheless, considering the development in Asir (Saudi Arabia), Hodeida in the Northern Yemen, and the transformation of Oman into an oil state, the future for the Southern Yemen is surely problematical.

REFERENCES

Guardian Special Report, 'The United Arab Emirates', 24 April 1975
The Middle East and North Africa 1974-75, 21st edition, Europa Publications
The Kingdom of Saudi Arabia: Facts and Figures, 1974
The Great Water Projects, Ministry of Information, 1971
'Saudi Arabia', a survey in *The Times*, 9 April 1974
Nihad Ghandri, *The Great Challenge*, Ministry of Information, 1968
Peter Mansfield, *The Middle East: A Political and Economic Survey*, 4th edition, Oxford University Press, 1972
W. B. Fisher, *The Middle East*, 6th edition, Methuen, 1971
United Nations Demographic Yearbook, 1967
J. I. Clarke and W. B. Fisher, *Populations of the Middle East and North Africa: A Geographical Approach*, University of London Press, 1972
Department of Information, Muscat, Oman, London 1972
The Middle East and North Africa 1975-76. A Survey and Reference Book, 22nd edition, Europa Publications
Michael Adams, *The Middle East: A Handbook*, Anthony Blond, 1971

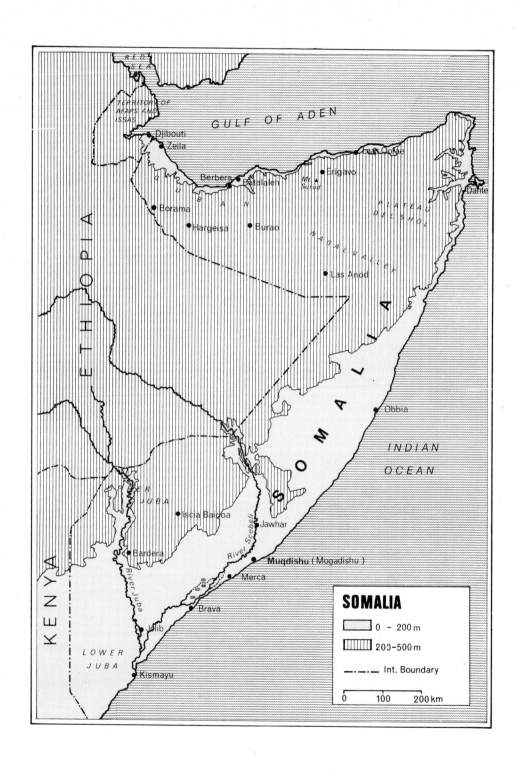

The Somali Democratic Republic

The Somali Democratic Republic, generally known as Somalia, is one of the latest members of the family of Arab nations, having joined the League of Arab States in 1973.

Somalia occupies most of the Horn of Africa, stretching as it does from the Red Sea in the north to beyond the Equator in the south and having an area of approximately 642,000 square kilometres. Its boundaries are formed to the north by the Gulf of Aden, to the east by the Indian Ocean, and to the south and west by Ethiopia and Kenya. As is obvious from a glance at the map, Somalia's location is of considerable strategic significance, the country constituting a focal point for the hinterland of Africa, the Indian Ocean, the Arabian Peninsula, and the Red Sea with the Gulf of Aden.

Somalia has a long, if only intermittently-recorded, history. This is the country known to the ancient Egyptians as the Land of Punt: between the seventh and the tenth centuries AD trading ports were established by immigrant Arabs and Persians along its extensive coastline; while occasional contact with traders from the Mediterranean lands may have occurred from Greek and Roman times. From the thirteenth to the fifteenth century there took place a series of Arab invasions from the Hadhramaut in southern Arabia, and it was during this period, too, that the hinterland was occupied by Somali tribes. The political and commercial importance of the capital, Muqdishu (sometimes less correctly known as Mogadishu or Mogadiscio) was established at about this time, while the Muslim Kingdom that existed here during the fifteenth and the sixteenth centuries waged war fairly continuously with neighbouring Ethiopia.

In 1889, during the main era of imperialist expansion into Africa, Italy established what became known as Italian Somaliland in the south of the country, while in 1895 Britain annexed the north, calling it British Somaliland. During the Second World War Britain occupied the south, displacing the Italian rule there, and this remained the situation until 1950, when the United Nations returned the southern part of the country to Italian administration as the Trust Territory of Somalia, with a view to the complete independence of both halves after a period of ten years. Thus on 1 July 1960 the two territories were at last reunited and together formed the independent sovereign Republic of Somalia.

In the years following her independence, Somalia faced a number of boundary problems involving disputes with her neighbours. Throughout history Somalia, in common with so many other countries, had not had formally fixed borders, nor, apart from the sea to the north and east, was there any obvious natural line of demarcation. Furthermore, the fiercely independent and martial Somali people are mostly nomads, to whom artificial lines drawn on a map mean little or nothing. As their whole traditional way of life involves constant movement in search of water and grazing-lands, and as other Somalis live on the other side of these artificial bounds, it is hardly surprising that difficulties arose. There have been, and to some extent there still are, problems to be solved between Somalia and Ethiopia, Kenya and the Somali Coast, otherwise known as the Territory of Afars and Issas, which is still under French control. However, by 1967 a border agreement had been signed with Kenya, and progress with these problems has undoubtedly been made.

A more fundamental and long-term problem for Somalia is, however, that of economic development. Her situation in this respect is not an easy one: her population now numbers between 2,500,000 and 3,000,000, of whom about 80 per cent are traditionally nomadic: she has few mineral resources, and most of the country is arid or semi-arid. To understand the extent of Somalia's task, it is necessary to look in more detail at the nature of the land and its resources, at what has been and is being done to develop it, and at future possibilities.

GENERAL DESCRIPTION

In the north and north-east of Somalia there are some hilly and mountainous areas, but in general the land consists of a flat, low, dry plateau. The two major permanent rivers lie in the southern half of the country: these are the Juba and the Scebeli, both of which rise in the highlands of Ethiopia.

Along the entire northern coast there runs a barren alluvial plain, called the Guban (literally, 'the burned'), where the atmosphere is both hot and humid, the rainfall low and the vegetation sparse. Frankincense and myrrh, however, are obtained here—in fact, the Islamic people of the Horn of Africa produce much of the incense used in Christian churches throughout the world.

This coastal plain varies considerably in width, widening from east to west, and gives way to a slightly more elevated inland plain with occasional limestone ridges and igneous rocks. Beyond this plain lies a limestone escarpment which constitutes the highest part of the whole of Somalia, rising to almost 2,450 metres at Mt Surud Ad, near Erigavo. From these mountains a huge, featureless plateau slopes to the south and south-east: here it is very dry and the vegetation consists of tall grass, thorn bushes and euphorbia trees. This plateau falls gradually into the Nagal Valley, whose name signifies 'the fertile land', but whose fertility has been diminished by severe and prolonged erosion.

The centre of Somalia consists mostly of low plateau and of plains fringed with coastal dunes, at no point rising to more than about 120 metres above sea-level. South-west of the River Juba, however, the bush becomes very thick, mangrove swamps are found, and wild life is abundant.

CLIMATE

In general, the southern half of Somalia is less arid than the north, owing to better rainfall and to the presence of the two main rivers. Taking the country (all of which lies in the tropical and sub-tropical zones) as a whole, the climate is fairly uniform throughout the year: the towns of Hargeisa, Borama, Burao and Erigavo in the north-west and north enjoy somewhat cooler weather than the rest of the country from October to February, and the temperatures have been known to drop to freezing-point on occasion. The coastal areas of the north are hotter, with a humid and enervating atmosphere—Berbera and Batalaleh endure temperatures of 42°C from June to September. In the south, temperatures are more uniform, with the coastal plains hot and wet and the hinterland hot and dry.

The year is divided in Somalia into four distinct seasons, two rainy and two dry. The first rainy season, called *gu*, lasts from March to June; the second, known as *dhair*, from September to December. The two dry seasons run from January to March (*jilal*) and from July to September (*hagai*). These seasons are not, of course, absolutely regular, and occasionally the expected rains fail to arrive. Drought is, not surprisingly, greatly dreaded by the Somali people.

Most of northern Somalia has under 250 mm of rain annually and many areas get less than 125 mm. Here most of the rain falls in the earlier of the two rainy seasons, the *gu*, whereas farther south the reverse occurs. In the south the rainfall is generally heavier than in the north: Muqdishu enjoys about 400 mm per year, and this can rise to 500 mm or more in the coastal areas of the south-west, diminishing farther inland. However, rainfall is uncertain, and drought, like that of 1964–65, can cause famine conditions; while conversely heavy rainfall can occur, causing flooding and ruining crops and plantations.

NATURAL RESOURCES

The main single natural resource of Somalia is its livestock: camels, cattle, sheep and goats. The camel is, as elsewhere, the mainstay of the nomad, who depends upon this indispensable beast for milk, hides and transport. Somalia's total livestock population is estimated at getting on for fifteen million head. About 80 per cent of the people are pastoral nomads and depend upon their domesticated animals, and there is an export trade in camels, sheep and cattle to the rest of the Arab World, particularly to Egypt and Saudi Arabia.

One third of the country is unsuitable for either agriculture or livestock, though half can be used for the breeding of livestock. The Livestock Development Agency is concerned with all aspects of animal production and marketing. Efforts are being made to improve the breeds of sheep on a more selective basis, and to develop poultry rearing (the government has set up a poultry farm in the Northern Region for the specific purposes of obtaining better strains of chicken).

The increase in cattle production is due to improvement in the control of diseases and of such pests as the tsetse fly, to increased feed, and to the development of new water supplies from wells and reservoirs.

In the industrial sphere, development is both recent and as yet very limited. No oil has yet been located in Somalia, but she does possess known mineral resources in the form of iron ore, tin, gypsum, bauxite and uranium ore. It is the last of these which offers the best chance of providing a base for speeding the economic growth of the country. A start has been made in developing her manufacturing and processing potential, with a major sugar factory at Jawhar, a rope-works at Merca, a

The cattle market at Burao.

Agriculture is moving ahead fast. Tobacco is now being cultivated successfully to save foreign exchange.

meat-packing station at Kismayu, and the largest fish-processing plant in East Africa at Laas Qorae.

AGRICULTURE

Cultivation in Somalia is severely restricted by the low average rainfall and by meagre resources of water, both on and below the surface. Nevertheless wells are frequent in the north in the mountains but scarce in the coastal belt. Ground-water in the south is inclined to be brackish. Settled agriculture is confined to a relatively few scattered areas, only about 14 per cent of the land being suitable for agricultural development. For rain-grown crops, the main centres lie along and between the rivers Juba and Scebeli, where the rains are more reliable as well as heavier than elsewhere. A United Nations report revealed that out of about eight million hectares of fertile land between the Juba and the Scebeli only 38,000 hectares were under cultivation. A further 800,000 hectares could, it is thought, be used for tillage between the two rivers.

Irrigation agriculture in the north is largely confined to date-palm plantations in the coastal plain, and in the south once again to areas along the two large rivers. The Scebeli is usually dry in February and March, but its average flow measures about 140 cubic metres per second. The flow of the Juba is two and a half times greater, and this river very rarely dries up entirely at any time.

The main crops in the southern cultivated lands are millet, sugar cane, bananas, sorghum, rice, maize, groundnuts, sesame oil and cotton. The costs of production of sugar are unfortunately too high at present to enable Somalia to compete in the world export market, and the only cash crop exported in quantity is bananas. These amount to about two-thirds of the country's exports by value, some $3\frac{1}{2}$ million stems being sent abroad each year. They were originally introduced by the Italians, and are grown along the main rivers. There are also some reserves of timber.

The production of cotton, rice and vegetables could be stepped up in the riverine lands if the irrigation facilities were greater. Nevertheless the potential land development figure has been estimated at some 7·5 million hectares. This could effect a radical change in the way of life of the people. Enough grain could be grown for home consumption and a surplus exported. At the present time experiments are being carried out in selected areas of irrigated land where animal foodstuff is being grown for the fattening of stock prior to export.

There are a number of schemes in train for the construction of dams and reservoirs and for growing other cash crops, such as citrus fruits; while inundation farming now takes place on some 20,000 hectares in Lower Juba. In the great northern plateau, too, the lack of rain has been combated to an extent by the construction of *barkado*, or artificial reservoirs, in order to supply the needs of the nomads and their livestock. These have given rise to permanent village settlements near their sites, to the diminution of the traditional competition between groups of nomads for water and grazing land, and to the introduction of a money-economy into the hinterland, where money has hitherto played little or no part in the lives of the inhabitants.

However, the expansion of water-conservation and of irrigation is inevitably slow, owing to shortage of money as well as the other difficulties already mentioned. To achieve economic viability (other than through finding oil) Somalia needs, and is receiving, aid from various sources. To enable her to face the continued heavy expenditure so necessary for the expansion of her cultivated lands and for the maximum exploitation of her natural resources of all kinds, the energies of her people are being mobilised in new ways by the government which came to power following the 1969 revolution. Crop diversification, for instance, is being practised on the State farms of Tugwajaleh in the north and Jelib in the south. Bank credit facilities are also available to assist with the efficient management of co-operatives, so that the small farmer with only two or three hectares of land can be brought into the technological age operating modern farming methods within the co-operative circle.

REFERENCES

The Middle East and North Africa 1969–70, 16th edition, Europa Publications
Somali Development Bank Report, Muqdishu 1971
Beautiful Somalia, Ministry of Information and National Guidance, Muqdishu 1972

Index

Index

El-Abala, Sudan, 94
Abbasi Canal, Egypt, 50
Abdel-Aal Power Plant, Lebanon, 199
Abha Dam, Saudi Arabia, 283
Abu Dhabi, 270, 280, 294
Abu Dibbis Basin, Iraq, 264–5
Abu Jebel Scheme, Sudan, 109
Abu-Zabad, Sudan, 95
Abyan, South Yemen, 304
Acre Plain, Palestine, 226
Adasiya, Syria, 240, 242–3
Aden, South Yemen, 303, 304
afforestation, 122–3, 131, 135, 138, 144, 154, 173, 178, 182, 193, 229, 292, 294–5
aflaj, 260, 295, 297
Aflaj, Saudi Arabia, 277–9
Afula, Palestine, 227
Agrarian Reform Authority, Egypt, 81
agricultural co-operatives, 77, 83, 130, 138, 178, 209, 217–19, 256, 259, 304
Agricultural Research Farm, Deirab, 279
agricultural self-administration, 158–62
Agricultural Trials Station, Digdagga, 294
agriculture, 62–5
   by burning, 104
   development of, 13, 15, 16
   mechanisation of, 77, 83, 104, 130, 193, 213, 218, 228, 280
   and population, 79, 81–3, 92–7
   sedentary, 105
   shifting, 103–5, 112, 301
   xerophilous, 98
   see also respective countries
Ahaggar Massif, Algeria, 156
Ahmadi Ridge, Kuwait, 286
Ahmed, Tewfik Hashim, 108
Al-Ain, Abu Dhabi, 294–5
Ait Aidel Dam, Morocco, 180
Ajman, 294
Akjoujt, Mauritania, 186
Aleppo, Syria, 201, 203, 204, 212–13
Alexandria, Egypt, 47, 55

Algeria, The Democratic Republic of, 16, 149–64
   agrarian reform in, 162–3
   agriculture, 151–62
   and Arabism, 19
   climate, 150
   coastal plain in, 149–51 *passim*, 152–5
   communications, 155
   crops grown, 152–3, 155, 158
   forests, 154–5
   French colonialism in, 151, 162–3
   irrigation, 152–3, 157
   land tenure, 151, 158–9, 162–3
   livestock in, 155, 156
   map, 148
   oases, 149, 150, 156–7
   oil revenue, 151
   population, 149, 152, 155
   self-administration in, 162–3
Algiers, 149, 152, 153, 155, 162
alluvial soils, 45, 64, 152, 155, 227, 304
Al-Amara, Iraq, 250, 258, 266
American Technical Aid Project, 138
Amlaj, Saudi Arabia, 274
Amman, Jordan, 230
Amr Ibn el-Aas, 21, 22
Animal Production and Agricultural Research Centre, Al-Hufuf, 279
Annaba, Algeria, 149, 152, 153
Ansariye Mts, Syria, 203–4, 214
Anti-Atlas Range, Morocco, 167, 169
Anti-Lebanon Range, 187, 201, 203
Aqaba, Gulf of, Jordan, 231
El-Arab R., Sudan, 95
Arab Company for Land Reclamation, 72
Arabia Felix, 301
Arabian Gulf, 16, 248, 270, 272, 274
Arabian Peninsula, 16, 270–1
Arabian Sea, 18
Arab–Israeli wars, 225; *see also* Suez War
Arab Lands, 13–19
   foreign domination of, 19
   inhabitants, 18–19

natural boundaries of, 13
Arab League, 182, 183, 242-3, 245-6, 307
Arabs, 18-19
Arbil, Iraq, 253
Arid Lands Research Centre, Abu Dhabi, 295
artesian wells, 71, 120, 130, 135, 153, 157, 204,
    232, 268, 282, 290
Asfun Canal, Egypt, 50
Al-Ashar Plain, Syria, 206
Al-Asharna Dam, Syria, 206
Asi R., see Orontes R.
Asir, Saudi Arabia, 272, 275, 277, 305
Aswan, Egypt, 30
Aswan Dam, 31, 34, 49-50, 55
Aswan High Dam, 23, 30, 34, 37-40, 42, 51, 56
    and crop production, 83
    and land reclamation, 59-61, 67, 72
    and Sudan, 103
Asyut, Egypt, 49
Asyut Barrage, 31-2, 49-50
Atar Oasis, Mauritania, 183
Atbara R., Ethiopia, 30, 90-2 passim, 109
Atlas Mts, Maghreb, 18, 149, 152, 155, 166, 167,
    169
Atmour Desert, Sudan, 89
El-Atrun Oasis, Sudan, 90
Awali R., Lebanon, 199
Azande tribe, 97

El-Baggara (cattlemen), 95, 97
Baghdad, Iraq, 252, 263, 264, 266, 267
El-Baguriya Canal, Egypt, 51
Bahrain, The State of, 270, 290
Bahr el-Ghazal Basin, Sudan, 28, 52, 91, 92, 97
Bahr Shibin Canal, Egypt, 50, 51
Bahr Yusef, Egypt, 22, 49
Baiyuda Desert, Sudan, 90
Bakhma Dam, Iraq, 265
Balfour Declaration 1917, 223, 224
Balikh R., Iraq, 259
Baniyas R., Syria, 237-8, 240-5
Barada R., Syria, 203, 211
Baraka R., Sudan, 98-9, 108
Baraka Delta Scheme, 108
barari, 45
Barbour, K. M., 107
Bardouni R., Lebanon, 200
Bareighit R., Lebanon, 237-8, 244-5
barrages, 31, 153, 179, 206, 208, 264-6, 304
Al-Basra, Iraq, 258, 267, 274
Batalaleh, Somalia, 309
El-Batana Plain, Sudan, 90
Batina Coast, Fujaira and Oman, 295, 297-8
Behariya Canal, Egypt, 47, 51
Behariya Oasis, 69

Beida, Libya, 115, 122
Beirut, Lebanon, 187, 198
Beit Hanum, Palestine, 237
Beja R., Tunisia, 134
Beja tribe, 94
Benghazi, Libya, 18, 115, 117
Beqaa, Lebanon, 187, 192-3, 198-9, 203
Berbera, Somalia, 309
Berbers, 152, 155, 166, 175
Birin Oasis, Saudi Arabia, 279
Bissan, Palestine, 240
Bitter Lakes Project, Egypt, 72
Bizerta, Tunisia, 146
Blue Nile R., 26, 28, 30-1, 90-2 passim, 97, 105
    dams on, 99, 101, 106
Borama, Somalia, 309
Al-Borma oil field, Tunisia, 134
Boufisha, Tunisia, 135, 138
Bougie, Algeria, 152, 155
Boutoun Basin, Israel, 243-4
Brakna, Mauritania, 183, 185
Britain
    in Iraq, 257
    and Palestine, 223, 224, 240
Buhiya Canal, Egypt, 51
Buraimi Oasis, Abu Dhabi, 294
Burao, Somalia, 309

Calanscio Desert, Libya, 120
calendar, 23
camels, 94, 298
canals, 263, 267
Casablanca, Morocco, 173, 175, 177
Cataract Hotel, Aswan, 56
cattle, 95, 97, 108
Cheliff Valley, Algeria, 152
Chemama, Mauritania, 185
civilisation, origins of, 13-15
Clawson, Marion, 131
coastal plains, 16, 18, 115, 117, 123, 149-51
    passim, 152-5, 172-3, 214, 226, 297
Collo Massif, Algeria, 152
Colomb Bechar, Algeria, 155
Col of Taza, Morocco, 172
cultivation, see agriculture; land reclamation
Cyrenaica, Libya, 117, 120, 121
Cyrene Plateau, Libya, 122

Ad-Dahna Desert, Saudi Arabia, 271, 274, 282
Dakhla Oasis, Egypt, 55, 69
Damascus, Syria, 201, 203, 211
Damietta Branch (Nile), 30, 45, 47, 51
    barrage on, 31
dams, 31, 45, 98-9, 131, 143, 157, 166, 178, 185,
    198-9, 230, 236, 243, 283, 304, 311

Dan, R., Syria, 237-8, 240-1
Ad-Darb, Saudi Arabia, 277
Darbandikhan Dam, Iraq, 252, 265
Darfur Province, Sudan, 90, 92, 95
    agriculture, 97, 104
    irrigation, 98
    land tenure, 111
Darik, Syria, 213
Dariya Dam, Saudi Arabia, 283
Dawson, Sir Ernest, 258
Dead Sea, Palestine, 229, 230, 238, 240, 242, 244
Deir el-Balah, Palestine, 229
Deir ez-Zor Province, Syria, 204
desalination of sea water, 52, 277, 287, 292, 295
deserts, 13, 270, 282
Dhibban, Iraq, 264
Dhofar Plateau, Oman, 297-8, 304
Dibbis Dam, Iraq, 253
dikakah, 275
Dinka tribe, 95, 97, 104
dira, 257
Diyala R., Iraq, 252, 265, 266
Djebilet Plain, Morocco, 167
Djerba Island, Tunisia, 139
Doha, Qatar, 292
Dokhan Dam, Iraq, 264, 265
Dongola Bend (Nile), 89
Douled oil field, Tunisia, 134
Draa Valley, Morocco, 168
drainage, 45, 52-3, 59, 64, 83, 143, 206, 207, 230,
    262, 267-8, 282
    aretic, 265
Dubai, 294
Dujaila, Iraq, 259
Dunqul, Sudan, 102

Eastern Desert, Jordan, 232
Eastern Highlands, Egypt, 79
East Ghor Canal, Jordan, 235, 236, 237, 242, 245
Egypt, The Arab Republic of, 21-86
    agriculture, 13, 21, 37, 55-6, 75, 77, 79
    ancient civilisation of, 13-15, 22-3
    climate, 21, 52
    crops grown, 45-7, 57, 75, 77, 78, 82-4
    fishing industry, 37
    geography, 21
    irrigation and drainage, 22, 31, 37-38, 42-53,
        55-6, 59, 72
    July Revolution 1952, 18, 56, 81
    landownership, 81-3
    land reclamation, 56-78
    livestock, 75-7, 84
    population, 79-84
    power sources, 37
    see also Nile Delta; Nile R.; Nile Valley

ein, 279
Ein Hunayn, Saudi Arabia, 277
Ein Zubaida, Saudi Arabia, 277
Equatoria Province, Sudan, 92
Erigavo, Somalia, 308, 309
Eritrean Plateau, 90
Eski-Kalek project, Iraq, 253
Esna Barrage, Egypt, 32, 50
Essaouira, Morocco, 172
Ethiopia, 308
Ethiopian Plateau, 91
Euphrates R., 13, 15, 16, 272
    barrages on, 264-6
    in Iraq, 248-9, 252, 263, 267
    seasonal flow of, 259, 261, 263
    in Syria, 201, 203, 205, 214, 217
Euphrates Dam, Syria, 206, 207, 208, 215, 219

Faisal Bedouin Resettlement Scheme, 280-1
Faiyum, Egypt, 22, 49, 53
Fallujah, Iraq, 264
Farafra Oasis, Egypt, 69
farmers, 97, 125-31
Faylakah Island, Kuwait, 286
Fderik, Mauritania, 186
Fertile Crescent, 201, 203
Fez, Morocco, 172, 177
Fezzan, Libya, 115, 120
First World War, 223
fishing, 37, 173, 183, 287, 290, 304
foggara, 157
Fujairah, 294, 295
Fur tribe, 97

Gafsa, Tunisia, 135
Galilee, Palestine, 226, 240-1
gas, liquid, 149
Gash Delta Scheme, Sudan, 107
Gash R., Sudan, 98-9
Gaza, Palestine, 226, 229
Gaza Plain, 226, 229
General Confederation of Algerian Trade Unions,
    160
General Desert Development Organisation, 67,
    68-72
General Egyptian Land Reclamation Organisa-
    tion, 67, 72-4
General Egyptian Organisation for the Exploita-
    tion and Development of Reclaimed Lands,
    67, 74-8
General Land Development Organisation, 67-8
General Organisation for Agrarian Reform and
    Rehabilitation of Land, Libya, 125-7
geodetic surveys, 60-1
Gezira, Syria, 201, 204, 207, 210, 213, 217

Gezira Plain, Sudan, 90, 97, 111
    agricultural scheme, 105-6
    irrigation, 98, 99
Al-Ghab project, Syria, 206, 212
ghaffar flies, 94
Gharbiya (Faroukia) Canal, Egypt, 50
Al-Gharraf, Iraq, 250, 266
ghibli, 120
Al-Ghor, Jordan, 227, 229, 230, 240, 245
Ghouta Oasis, Syria, 203
Golan Heights, Syria, 211
Gorgol valleys, Mauritania, 185
Great Rift Valley, 90
Great Syrian Desert, 232
Guban, Somalia, 308

haboob, 99
habous land, 145, 146
hadhabat, 275
Hadramaut, South Yemen, 271, 304-5, 307
hafir, 98
Haiel, Saudi Arabia, 277
Al-Haiet Oasis, Saudi Arabia, 274
Haifa, Palestine, 226
Halba, Lebanon, 192
Hama, Syria, 203, 206, 212-13
hamad, 115, 117
Hamad, Saudi Arabia, 272
Hammadat, Morocco, 176
Hanakia Oasis, Saudi Arabia, 274
Haouz Plain, Morocco, 167, 174, 180
Hard, Saudi Arabia, 280
Hargeisa, Somalia, 309
harras, 271
Al-Hasa, Saudi Arabia, 272, 274-5, 277, 279, 282
Hasbani R., Lebanon, 197-8, 237-8, 240-6
El-Haseke Province, Syria, 203, 204, 207
Hassan Addakhil Dam, Morocco, 180
Heet, Saudi Arabia, 277
herdsmen, 94-7, 157
Herodotus, 21, 23
Herzl, Theodor, 223
High Atlas, Morocco, 167, 169
High Plateau of Shotts, Algeria, 149-50, 151, 155-6
High Tell, Tunisia, 134-5
Hijaz, Saudi Arabia, 273-5 *passim*, 277
Hilla, Iraq, 266
Hindiya Barrage, Iraq, 264, 266
Hodeida, North Yemen, 302, 305
Hodh, Mauritania, 183
Homs, Syria, 203, 206, 212-13
Houran Plain, Syria, 203, 210-11
Hufuf, Saudi Arabia, 274, 280
Huran, Saudi Arabia, 272

Hurst, H. E., 103
Huseiniya Plain, Egypt, 78
Hussein, Sheikh, 223
Al-Huwait Oasis, Saudi Arabia, 274
Huwar Island, Bahrain, 290
hydro-electric power, 37, 169, 199, 206, 207, 238, 240-3, 265

Ibn Batuta, 18
Ibn Khaldun, 18
Ibrahimiya Canal, Egypt, 32, 49
Idfina, Egypt, 50-1
Indian Ocean, 275
Institute of National Planning, Egypt, 77-8
International Institute for Wheat Studies, Mexico, 279
Iraq, The Republic of, 207, 248-68
    agriculture, 248-56
    crops grown, 249-50, 252-3, 262
    land tenure, 257-9
    maps, 251, 260
    population, 248
    rainfall and irrigation, 248, 252-3, 263-8
    river flood plains, 13, 15, 16
    water resources, 259-62
Ironstone Plateau, Sudan, 92
irrigation, 15
    basin, 42-7, 49-50, 98, 102
    canal, 99-101, 109, 178-9
    by *dalu*, 115
    by inundation, 98-9, 115, 304-5, 311
    and land reclamation, 78
    perennial, 42, 47-53, 59, 72, 83, 115
    pump, 101-2, 109, 130, 142, 206, 207, 268
    by sprinkling, 75
    see also respective countries
Ishaqi, Iraq, 267
Ismail, Khedive, 56
Ismailiya Canal, Egypt, 47, 49, 51
Israel, 223, 225
    and Jordan River projects, 241-6

Jaffa, Palestine, 226
Al-Jalil Plateau, Palestine, 227
Jawhar, Somalia, 309
Jebel, Libya, 115
El-Jebel R., Sudan, 97
Jebel al-Akhdar, Libya, 18, 117, 122-3, 130-1
Jebel al-Akhdar, Oman, 270, 297
Jebel Aulia Dam, Egypt, 31, 32, 34, 50
Jebel Druze, Syria, 201, 210-11
Jebel Dukhan, Qatar, 290
Jebel Marra, Sudan, 89-90
Jebel Nafusa, Libya, 115
Jebel esh-Sheikh, Syria, 203-4, 230, 237

317

Jebel el-Taf, Saudi Arabia, 274
Jeddah, Saudi Arabia, 270, 274, 277
Jefara, Libya, 115, 120–3 *passim*
Jelib, Somalia, 312
Jenin, Palestine, 227
Jerusalem, 229
Jezril, Vale of, Palestine, 227
Jiddah Island, Bahrain, 290
Johnston, Alec, 241, 243
Johnston Plan, 241–2
Jordan, The Hashemite Kingdom of, 208, 226,
    230–46
  agriculture, 236–7
  climate, 230
  crops grown, 230, 235–6
  irrigation, 230, 236, 238, 244–6
  land tenure, 232–5
  map, 231
  oases, 232
Jordan R., 16, 198, 203, 208, 226, 230, 232
  average yearly flow of, 238, 241
  projects, 237–46
Jordanian Plateau, 230–2
Joseph, 16, 25
Jounie, Lebanon, 199
Juba R., Somalia, 308, 311
Judaean Plateau, Palestine, 227, 229

Kababish tribe, 94
Kabir Organisation, Algeria, 160
Kabylie, Algeria, 152, 155, 159
El-Kadi Hills, Syria, 237, 238
El-Kafir R., Syria, 206
Kagera R., Tanzania, 26
Kairuan, Tunisia, 141
Karaoun, Lebanon, 199
Karbala, Iraq, 266
Kasr Attra Dam, Israel, 243
Kassala, Sudan, 90, 92, 98, 107, 111–12
Kassala Cotton Company, 105, 107
Kedia d'Idjil Hills, Mauritania, 186
Kelabiya Canal, Egypt, 50
Kenitra, Morocco, 173
Kenya, 87, 308
Khabur R., Syria, 203, 205–7, 214, 259, 261
Khabura Desert, Saudi Arabia, 274
El-Khalil (Hebron), Palestine, 229
Kharma Range, Saudi Arabia, 274
khamsin, 21, 22
Khan Yunis, Palestine, 229
Khardala, Lebanon, 198
Kharga Oasis, Egypt, 55, 69
Al-Kharj, Saudi Arabia, 272, 277–9
Khartoum, Sudan, 26, 28, 30, 92, 98, 99
  climate, 87

Khashm el-Girba, Sudan, 109
Khibar, Saudi Arabia, 274, 277
Kirkuk, Iraq, 253, 267
Kismayu, Somalia, 311
Kordofan Province, Sudan, 90–2 *passim*, 95
  agriculture, 97, 104, 109
  land tenure, 111
Kroumirie Mts, Tunisia, 134–5
Ksar el-Kebir, Morocco, 175
ksout fortress, 168
Kufrah Oasis, Libya, 120, 130
Kut, Iraq, 250, 252, 263
Kut Barrage, 266
Kuwait, The State of, 248, 271, 286–7
  map, 285
Kuwait Fund for Arab Economic Development,
    287, 302

Laas Qorae, Somalia, 311
Lahej, South Yemen, 304
Lake Albert, Uganda, 26, 28, 30
Lake Edward, Uganda, 26
Lake Habbaniya, Iraq, 264–5
Lake Hammar, Iraq, 265, 267
Lake Homs, Syria, 208, 212
Lake Hula, Palestine, 232, 237–8, 241, 243
Lake Kyoga, Uganda, 26
Lake Manzala, Egypt, 51, 79
Lake Nasser, Egypt, 30, 37, 38, 71
Lake Tana, Ethiopia, 28
Lake Tiberias, Palestine, 229, 230, 238, 241–4
Lake Victoria, 26, 30, 51
landownership, *see* various countries
land reclamation
  and cultivation, 62–5, 75
  organisation of, 67–78
  process of, 59–62
  see also Egypt; Libya; Syria
Land Reclamation Organisation, Libya, 131
Larache, Morocco, 175
Latakia, Syria, 203, 204, 214, 217
Lawrence, T. E., 15
Lebanon, The Republic of, 187–200
  agriculture, 192–7
  climate, 192
  coastal plains, 18, 187, 192
  crops grown, 192–3, 196–7
  development plans, 196–200, 245
  irrigation, 192–3, 197–200
  land tenure and use, 196
  maps, 190–1
Lebanon Mts, 187, 193, 199, 227
Libya, The Arab Republic of, 114–33
  agriculture, 114, 121–31
  climate, 115, 120

318

crops grown, 121, 130, 132
  geography, 114-20
  irrigation, 17, 115, 117, 130
  land reclamation, 130-1
  land tenure, 131
  livestock, 132
  population, 114-15
  rural settlement, 125-9
Libyan National Agricultural Bank, 130
Litani R., Lebanon, 187, 192-3, 197-8
  development of, 199-200, 240
Lith, Saudi Arabia, 277
livestock, 75-7, 93-7, 111-12; see also various
  countries
Lord Milk Project, Jordan Valley, 240
Ludwig, Emil, 23

Maghreb, 16, 166
Maharès, Tunisia, 135
Mahmudiya Canal, Egypt, 47, 51, 55
Mahreda Dam, Syria, 206
Al-Makba R., Palestine, 226
Malakal, Sudan, 28
Malaki Dam, Saudi Arabia, 283
Mameluke Sultans, 47, 55
Managil Scheme, Sudan, 105, 106-7
Manamah, Bahrain, 290
Maqarin, Syria, 240-1, 243
Al-Maqrizi, 55
Marib Dam, Yemen, 13, 15
Maridi, Sudan, 109
marigots, 185
Marjayoun Project, Lebanon, 197
Markabe Tunnel, Lebanon, 199
Marsa Matruh, Egypt, 55
Maryat Project, Egypt, 72
Mateur, Tunisia, 135
Matmata Hills, Tunisia, 142
Mauritania, The Islamic Republic of, 182-6
  agriculture, 183-6
  climate, 182
  crops grown, 185
  fishing industry, 183
  industry, 183, 186
  livestock, 183
  population, 182-3
Mecca, Saudi Arabia, 15, 16, 270, 272, 274, 277
Medina, Saudi Arabia, 270, 274
Medjerda R., Tunisia, 134, 138, 143
Medjerda Valley Scheme, 143
Meknès, Morocco, 172, 177
Melleque R., Tunisia, 134, 143
Menzel Bourguiba, Tunisia, 135, 138
Menzel Temime, Tunisia, 135
Merca, Somalia, 309

Merj Ibn-Amer, Palestine, 227, 241
Mernak, Tunisia, 138
Meseta Plateau, Morocco, 167, 174
meska, 142
Mesopotamia, 13, 15, 16, 248
Mezerili, Syria, 240
Middle Atlas, Morocco, 166-7, 169
Milk, Walter Clay Lord, 240
Mina al-Ahmadi, Kuwait, 287
Minufiya Canal, Egypt, 47, 51
Minya Project, Egypt, 72
Mocha, North Yemen, 301
Mogod Mts, Tunisia, 134
Mohammad Ali, 31, 47, 55
Mokhiba Dam, Jordan, 246
morabaa, 209
Morocco, The Kingdom of, 16, 166-80
  agriculture, 171-6
  climate, 169
  coastal plains of, 18, 172-3
  crops grown, 168, 171-5, 178
  fishing industry, 173
  French colonialism in, 177
  geography, 166-9
  irrigation, 166, 169, 171-4 passim, 178-80
  land tenure, 177-8
  livestock, 174, 176
  map, 169
  oases, 168, 176
  phosphates, 178, 180
  population, 166, 178
moshanka, 210, 213, 214
Mosul, Iraq, 253, 265, 267
Moulouya R., Morocco, 166, 168, 175
Mount Carmel, Palestine, 226
Mount Hermon, see Jebel esh-Sheikh
Mount Surud Ad, Somalia, 308
Muharraq Island, Bahrain, 290
Mujamma Dam, Saudi Arabia, 283
Mujara, Iraq, 265
Mukalla, South Yemen, 304
Muqdishu, Somalia, 307, 309
Murchison Falls, Uganda, 26
Murzuq Basin, Libya, 131
Musayyib, Iraq, 252, 259, 265
Muscat, Oman, 270, 296
Al-Muwaylih, Saudi Arabia, 274

Nabeul, Tunisia, 135, 138
Nabih Salih Island, Bahrain, 290
Naccache Dam, Lebanon, 199
Nafud Desert, Saudi Arabia, 270, 271, 282, 284
Nagal Valley, Somalia, 308
Nag Hammadi Barrage, Egypt, 32, 50
Nahrwan Canal, Iraq, 263

Najran, Saudi Arabia, 275, 277
Napoleon Bonaparte, 15, 31, 47, 55
Nasiriya, Iraq, 250
Nasser Canal, Egypt, 78
National Office for the Development of the Socialist Sector, Algeria, 160
National Organisation for Agrarian Reform, Algeria, 160, 162
National Organisation for Agricultural Settlement, Libya, 125
Negev, Palestine, 226, 227, 229, 238, 240, 243, 246
Nejd, Saudi Arabia, 272, 275, 277
New Valley Project, Egypt, 68–71
Nile Delta, 18, 21, 30–1
  barrages in, 31, 35, 47, 50–1, 55, 78
  drainage, 53
  irrigation, 45–51
  population, 79
  soils of, 64
Nile R., 13, 15, 16, 21, 22–41, 259
  cataracts, 30
  course of, 26–31
  flood, 22–3, 25, 30, 45
  and irrigation, 22, 31, 37–41, 42–53
  reservoirs on, 31–38
  sources of, 22
  in Sudan, 89, 91
Nile Valley, 13, 21
  and irrigation, 45, 59
  land reclamation in, 72
  Lower, 30
  population, 79, 97
  soils of, 64
Nile Waters Agreement, 52
nomads, 16, 90, 93–7, 111, 115, 123, 141, 182–3, 232, 272, 308, 309
  settlement of, 127–9, 142, 147, 280–1, 295, 311
norias, 206
Northern Plains, Syria, 213
Northern Province, Sudan, 98, 101, 111
Northern Tell, Tunisia, 134–5
North-Western Coast, Egypt, 71
Nouakchott, Mauritania, 182
Nuba Mts., Sudan, 90–1
Nuba tribe, 91, 97
Nubariya Canal, Egypt, 56, 78
Nubia, Sudan, 89
Nuer tribe, 95, 97, 104
Nukhaila Oasis, Sudan, 90
Nyala, Sudan, 95

oases, 13, 15, 16, 135, 294; see also various countries
oil, 114–15, 121, 134, 149, 151, 248, 270, 272, 284, 286–7, 290, 292, 298
Oliver, J., 92
Oman, The Sultanate of, 270, 296–8, 305
Oran, Algeria, 149, 152, 153, 155, 162
Organisation Commune des Régions Sahariennes (OCRS), 183
Organisation for the Exploitation of the Waters of the River Jordan and its Tributaries, 246
Orontes R., Lebanon, 16, 187, 192, 203, 205, 212
  Upper Orontes Project, 208
oroush land, 145, 146–7
Ottoman Government, 257, 263–4
Al-Ouja R., Palestine, 226
Oum er-Rbia R., Morocco, 166, 173
Owen Falls Dam, Uganda, 26

Palestine, 222–9
  climate, 226–9
  coastal plain, 226
  crops grown, 226–9
  geography, 226–9
  history, 222–5
  map, 228
  population, 224, 229
  refugees, 225, 229, 236–7
  Western Plateau, 227
Palestine Mts, 227
Port Etienne, Mauritania, 183, 186
Port Said, Egypt, 78
Port Sudan, 107
Prah Tekvah, Palestine, 226

Qamishliye, Syria, 204
Qasim, Saudi Arabia, 277
Qasimiya Project, Lebanon, 197, 199
Qatar, The State of, 270, 290–2
Qatif Oasis, Saudi Arabia, 274, 282
Qena Province, Egypt, 50
Qishon R., Palestine, 227
Qoz region, Sudan, 90
Quran, 15–16
Qurna, Iraq, 248, 249–50, 264

Rabat, Morocco, 171–3 passim
Rabigh, Saudi Arabia, 274
Rafah, Palestine, 229
Rafraf, Tunisia, 138
Rahad Project, Sudan, 101
Rakad Valley, Jordan, 245
Ramadi, Iraq, 252, 264, 266
Raqqa Province, Syria, 203, 204
Ras al-Khaimah, 294
Ras el-Tib Peninsula, Tunisia, 138
Al-Rastin Dam, Syria, 206
Raudhatain, Kuwait, 287

Red Sea, 47, 90, 270–2 *passim*
  coastal plains of, 18, 275, 299
Regreg R., Morocco, 173
Rehamna Plain, Morocco, 167
reservoirs, 31–38, 143–4, 199, 227, 264, 309, 311
Rharb Plain, Morocco, 167
Rif Atlas, Morocco, 166–7, 175
Rio de Oro R., Mauritania, 185
Riyadh, Saudi Arabia, 272, 280
Roseires Dam, Sudan, 101, 105, 106
Rosetta branch (Nile), 30, 45, 56
  barrage on, 31, 50–1
Rubin R., Palestine, 226
Rub al-Khali, Saudi Arabia, 270, 279, 282, 284
rural development societies, 77

Sadiyat Island, Abu Dhabi, 295
Al-Safra Oasis, Saudi Arabia, 274
Sahara Desert, 89, 94, 115, 135, 150, 166
Saharan Atlas, Algeria, 150, 155, 156
Sahel, Tunisia, 135
Saïs Plain, Morocco, 172
sakia (water-wheel), 46, 98, 102, 268
salinity, 64, 74, 83, 143, 262–3, 267, 287
Salum, Egypt, 71
Samarian Plateau, Palestine, 227
Samarra, Iraq, 252, 263, 265
Samawa, Iraq, 252
Sami, Amin, 25
Sanaa, North Yemen, 299
saniya, 117
Saudi Arabia, The Kingdom of, 270, 272–84
  agriculture, 279–82
  climate, 279
  crops grown, 275
  dams, 283
  geology, 275
  irrigation and drainage, 282–3
  land tenure, 281
  livestock, 279, 280–1
  oases, 274
  oil revenue, 284
  population, 272
  water resources, 274, 275–9
Scebeli R., Somalia, 308, 311
sebkha, sabkhat, 182, 275, 282, 292, 294, 301
Sebou R., Morocco, 167, 171–2, 178
seil flood water, 304–5
Senegal R., 182, 185
Sengar, Iraq, 253
Sennar Dam, Sudan, 32, 99, 101, 105–6
Sfax, Tunisia, 139, 142, 146
shadouf, 46, 98, 102, 268
Shamiya, Iraq, 265
Sharjah, 294

Sharqiya (Fouadiya) Canal, Egypt, 50, 51
Sharon Plain, Palestine, 226
Ash-Shatra drainage project, Iraq, 250
Shatt al-Arab, Iraq, 248, 249, 267, 287
Shatt al-Dujaila Canal, Iraq, 266
Shatt al-Halma, Iraq, 252
Shatt al-Hindiya, Iraq, 252
Shawia Plain, Morocco, 177
Shaykhan, Iraq, 253
Sheikh Said, Syria, 207
Ash-Shihr, South Yemen, 304
Shilluk tribe, 95–7, 104
Al-Shiqaya, Kuwait, 287
Shott Djerid, Tunisia, 135
Shott el-Hodna, Algeria, 150
shotts, 135, 149, 155
Shuwaykh, Kuwait, 287
Sidi-bel-Abbes, Algeria, 154
Sidi el-Mesri, Libya, 130
Sidon, Lebanon, 192, 198, 199
Siliana R., Tunisia, 134
Sinai, Egypt, 18, 55, 71–2, 79
Sirte, Libya, 18, 122
Sitra Island, Bahrain, 290
Siwa Oasis, Egypt, 55, 69
Sobat R., Sudan, 28, 30, 91, 95, 97
Socotra Island, South Yemen, 303
soil research, 62, 64, 75
Soliman, Tunisia, 138
Somali Democratic Republic, The, 307–12
  agriculture, 311–12
  climate, 308–9
  coastal plains, 18, 311
  crops grown, 311
  geography, 308
  history, 307
  livestock, 309
  map, 306
  resources, 309–11
Somali Coast, 308
Sousse, Tunisia, 139, 146
Sudan, The Democratic Republic of the, 25, 52, 87–113
  agriculture, 92–4, 95, 97, 103–9
  climate, 87
  crops grown, 87, 92–3, 101, 105, 109
  dams, 32
  geography, 89–91
  irrigation and water resources, 93, 98–109, 305
  land tenure, 105–7 *passim*, 110–11
  livestock, 93–7, 111
  population, 87, 90–1, 93–7, 103–9
  soil, 92
Sudan Plantation Syndicate, 105
Sudd, the, Sudan, 28, 30, 92

Suez Canal, 49, 56, 303
Suez War 1956, 37
Sulaimaniya, Iraq, 265
Sulaiyil, Saudi Arabia, 277
Al-Summan Desert, Saudi Arabia, 274
Sykes–Picot Agreement 1916, 223
Syria, The Arab Republic of, 201–21
   agricultural co-operatives, 217–19, 220
   Agricultural Reform Law, 215, 220
   agriculture, 210–21
   climate, 201
   coastal plains, 214
   crops grown, 203–4, 206, 210–11, 213–14, 218–19
   irrigation and rainfall, 17, 203–8, 212, 214–15, 245
   land tenure, 208–10, 215, 219–20
   map, 202
   oases, 213
   phosphate industry, 213
   population, 201, 204, 220–1
   river flood plains, 16

tabu land, 258
Tadla Plain, Morocco, 167
Tadmoor Basins, Syria, 213
Tahrir District, Egypt, 37, 56–9
Taif, Saudi Arabia, 274, 277
Taizz, North Yemen, 301, 302
Tartus, Syria, 214
Tel Aviv, Israel, 226
Tel Hej, Palestine, 241–2
Tell Atlas, Algeria, 149–50, 152, 155
Tel Manas, Syria, 207
Tensift R., Morocco, 173
terus, 102
Tessa R., Tunisia, 134
Tetuan, Morocco, 175
Tewfik Canal, Egypt, 47, 50, 51
Tharthar Reservoir, Iraq, 264–5
Tiberias–Negev conduit, Israel, 229, 244
Tibesti Mts, Libya, 115, 120
Tigris R., 13, 15, 16, 201, 248–9, 252–3, 263–4, 267
   barrages on, 265–6
   seasonal flow of, 259, 261, 263
   sediment in, 259
Tihama, Saudi Arabia and Yemen, 270, 274, 275, 299, 302
Todra Valley, Morocco, 168
Tokar, Sudan, 98, 107–8, 111
topographical survey, 61
Tozeur Oasis, Tunisia, 141
Trarza, Mauritania, 183, 185
Tripoli, Lebanon, 192

Tripoli, Libya, 115, 117
Tripolitania, Libya, 115–17, 121
tsetse flies, 94, 95, 104, 309
Tuareg, 182
Tugwajaleh, Somalia, 312
Tunis, 135, 138, 146
Tunisia, The Republic of, 16, 134–47
   agriculture, 138–44
   climate, 134–5
   crops grown, 138–42, 144
   French colonialism in, 145–7
   geography, 134–5
   land tenure, 139, 145–7
   livestock, 135, 138, 141, 144
   oases, 141–2
   population, 134, 145
   water resources and irrigation, 139–44
Turkey, 207, 248
Tuwaiq Mts, Saudi Arabia, 279
Tuz Khurmatu, Iraq, 253
Tyre, Lebanon, 192, 198

Ula, Saudi Arabia, 274, 275
Umm an-Nassan Island, Bahrain, 290
Umm al-Qaiwain, 294
United Arab Emirates, 270, 294–5
   map, 284
United Nations Development Programme, 178
United Nations Food and Agriculture Organisation, 71, 123, 127–8, 138, 142, 279, 292, 304
United Nations Relief and Works Agency, 225, 236
United Nations Special Fund, 71
University College of North Wales, 279
Uniza, Saudi Arabia, 277
Upper Egypt, 45, 52–3
Uwayrid Range, Saudi Arabia, 274

Wadi Ajal, Libya, 130
Wadi Allaqi, Sudan, 90
Wadi Beersheba, Palestine, 231
Wadi Bisha, Saudi Arabia, 277
Wadi Dawasir, Saudi Arabia, 277
Wadi Faruq, Saudi Arabia, 274
Wadi Fatma Oasis, Saudi Arabia, 274, 277
Wadi Gabgaba, Sudan, 90
Wadi Halfa, Sudan, 87, 89, 102, 103
Wadi Hanifa, Saudi Arabia, 283
Wadi Haum, Saudi Arabia, 277
Wadi Ibra, Sudan, 90
Wadi Jazl, Saudi Arabia, 277
Wadi Jizan, Saudi Arabia, 283
Wadi el-Miah, Saudi Arabia, 274
Wadi el-Milk, Sudan, 90
Wadi Muqaddam, Sudan, 90

Wadi Natrun, Egypt, 56, 59, 71
Wadi Rama, Saudi Arabia, 277
Wadi Shati, Libya, 130
Wadi Sirhan, Saudi Arabia, 272
Wadi Traghan, Libya, 130
Wadi Zabid, North Yemen, 302
Wajh, Saudi Arabia, 274
wakf, 145, 177, 301
Warrar channel, Iraq, 264
water, 15–16, 74, 75; *see also* irrigation
Wazani–Baniyas–Yarmuk Canal, 246
Wazani Spring, Jordan, 245
weirs, 31, 266
Western Desert, Egypt and Sudan, 55, 61, 71, 79, 90
White Nile R., Sudan, 26, 28, 30–1, 90, 92, 95, 97
    Basin, 28, 98
    pumping schemes, 108
Willcocks, Sir William, 264, 266

Yambio, Sudan, 87, 109

Yanbual, Saudi Arabia, 274
Yarmuk R., Syria, 203, 208, 230, 236, 238
    projects, 240–3, 245
Yemen, 13, 15, 270–1
    Arab Republic of, 299–302
    irrigation, 98, 302, 304–5
    maps, 299, 303
    People's Democratic Republic of, 303–5

Zab Rivers, Iraq, 253, 265, 267
Zagros Mts, 204, 248
Zahle, Lebanon, 192, 200
Zande Scheme, Sudan, 105, 108–9
Zarqa R., Jordan, 238, 240
Zargha R., Tunisia, 134
Zarzis, Tunisia, 142
Zifta Barrage, Egypt, 31, 50
Zionist Movement, 222–3, 224, 240
Ziz Valley, Morocco, 168, 180
El-Zor, Jordan, 230

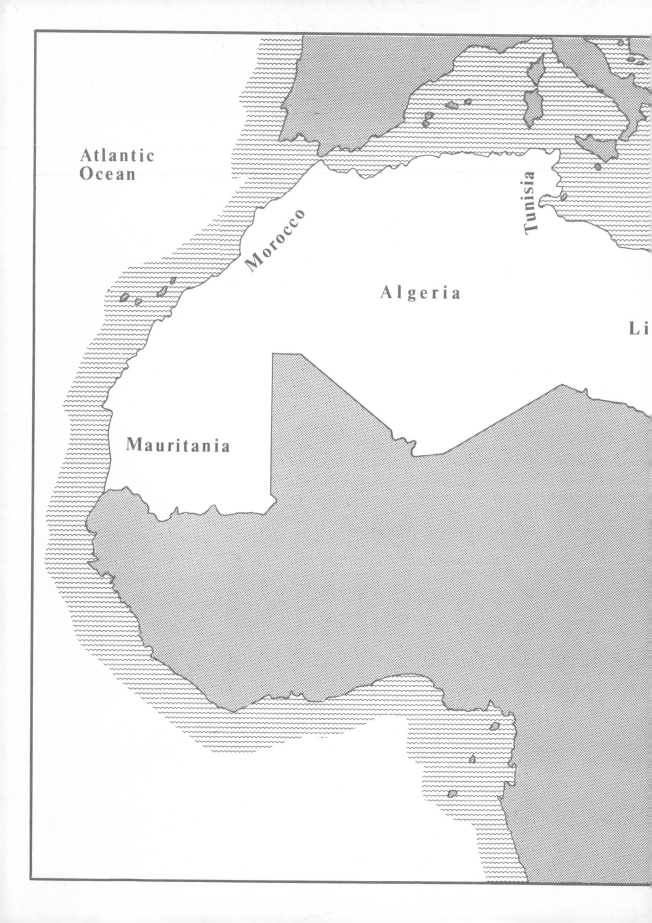